Digital Ethics

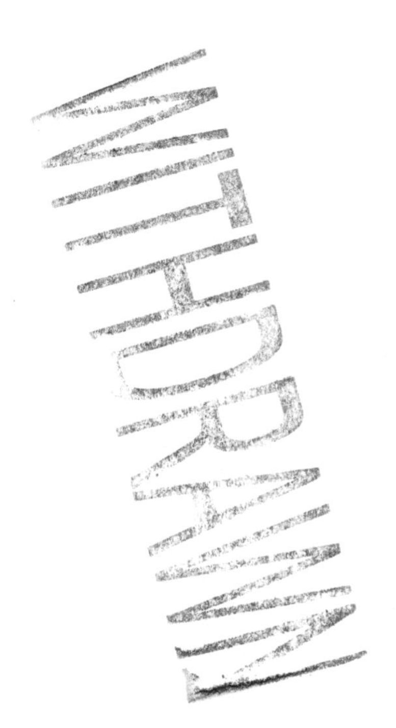

Steve Jones
General Editor

Vol. 85

The Digital Formations series is part of the
Peter Lang Media and Communication list.
Every volume is peer reviewed and meets
the highest quality standards for content and production.

PETER LANG
New York • Washington, D.C./Baltimore • Bern
Frankfurt • Berlin • Brussels • Vienna • Oxford

Digital Ethics

research & practice

EDITED BY DON HEIDER & ADRIENNE L. MASSANARI

PETER LANG
New York • Washington, D.C./Baltimore • Bern
Frankfurt • Berlin • Brussels • Vienna • Oxford

Library of Congress Cataloging-in-Publication Data

Digital ethics: research and practice / edited by Don Heider, Adrienne L. Massanari.
p. cm. — (Digital formations; v. 85)
Includes bibliographical references.
1. Internet—Moral and ethical aspects. 2. Internet—Social aspects.
3. Internet users—Psychology. I. Heider, Don. II. Massanari, Adrienne.
HM851.D527 302.23'1—dc23 2012032551
ISBN 978-1-4331-1896-8 (hardcover)
ISBN 978-1-4331-1895-1 (paperback)
ISBN 978-1-4539-0922-5 (e-book)
ISSN 1526-3169

Bibliographic information published by **Die Deutsche Nationalbibliothek.**
Die Deutsche Nationalbibliothek lists this publication in the "Deutsche Nationalbibliografie"; detailed bibliographic data is available on the Internet at http://dnb.d-nb.de/.

The paper in this book meets the guidelines for permanence and durability of the Committee on Production Guidelines for Book Longevity of the Council of Library Resources.

29 Broadway, 18th floor, New York, NY 10006
www.peterlang.com

Printed in the United States of America

Contents

Preface

Don Heider

I was in a role-playing community in a virtual world observing and trying to better understand what it meant to be a part of such a place. A woman who had been a long-time community member and who held a position of status and some authority in the community logged in one day with a new avatar. Instead of the petite redheaded female, she was now an athletic-looking male. She revealed to community members that she was in real life a man, which meant that a number of things she had told myself and others for months about her real life were fabrications. Reaction in the community varied widely. I was relatively new in the online world and had not encountered this kind of behavior before, but it seemed a clear violation of people's trust, a breach of ethics. Some in the community felt as I did; others did not. Almost every individual in the community had a slightly different take on the situation. I searched online for some kind of guidelines to ethical behavior in virtual communities, perhaps a code of ethics or best practices. None existed.

Many of us have had some similar experiences in this new digitized world in which we live. It might have been something as simple as opening an e-mail that gave your computer a virus or discovering that a blogger you trusted was actually being paid to blog about a particular product. Or you may have read in the news about a prominent person you admire texting a picture of his or her private parts to an unsuspecting recipient.

Since humans began speaking and interacting, we could argue ethical questions have existed. "Be so true to thyself as thou not be false to others," Francis Bacon wrote in *Of Wisdom for a Man's Self.* In an age of avatars, perceived anonymity, and diminishing face-to-face contact, what does it mean to be true to thyself? Has the Internet given us license to be false to others, without consequence? Ethics is a system or guide by which we ask questions about human behavior. Technology gave us capabilities we previously did not have. Technology changed the way we think about time and space. Internet technology brought with it new important questions and concerns.

As these ideas were at the top of my mind, I joined Loyola University in Chicago. The university itself had a rich tradition of engaging questions of ethics, growing out of its Jesuit roots. I found at Loyola some like-minded colleagues and the Center for Digital Ethics and Policy was born. We quickly realized the value in gathering an even larger group of scholars and thinkers to discuss these issues. That resulted in the first International Symposium on Digital Ethics convened at Loyola University. Authors and scholars from five countries came to participate and share ideas.

The conference resulted in a wonderful exchange of ideas, many of which are contained in this volume, such as questions about how we conduct research in a digital age, questions about citizen journalism, questions about disclosure in blogs, and much more. We hope this volume contributes to the growing body of knowledge in this new area of ethics. The four organizers of the conference offer brief section introductions, and the book begins with an overview of the field and this volume by one of the pioneers of the digital ethics field: Charles Ess.

Foreword

Charles Ess

Digital Ethics Past, Present, Futures

In the following, I offer an overview of Digital Ethics (DE) as a field of enquiry, which serves as the larger context for this volume and its constituent chapters. Along the way, I point out the trajectories and contributions of this volume and its constituent chapters to DE. To do so, I first take up four difficulties in facing efforts to develop a working definition of "digital ethics." I then offer a taxonomy of the ethical issues that constitute much of the main foci of DE over the past 5 years and indicate where these issues are further addressed in this volume. I conclude with some suggestions of emerging ethical domains and challenges that will likely become increasingly important components of DE, including one represented in this volume.

"Digital" Ethics?

There are at least four sorts of challenges to developing a working definition of DE. First of all, the term *digital* is itself both helpful and misleading. On the one hand, we intend the term digital to denote defining characteristics and affordances of specific technologies, as these lead to and illuminate a now very extensive range of ethical issues. Three examples of such characteristics can be noted here, beginning with how digitizing information—whether as originally sound, sight, or text—allows for the *convergence* of media forms, as exemplified in a contemporary smartphone that can record sound, take photographs, receive and send e-mail, and, oh yes, make a phone call. As well, in James Moor's (1985) term, digital information is "greased": it slides all but effortlessly and instantly through our networks and often to places and persons we do not intend. These features of digitized information catalyze many of our ethical concerns, beginning with *privacy* issues: What sorts of information about us *ought* we keep from flowing out from our devices—including, increasingly, the multiple kinds of information about our private lives that are documented and recorded by our smartphones which, by way of their portability, thus reach into more or less every nook and cranny of our lives? Finally, these privacy concerns are radically complicated and extended as digital information and media channels are distributed through a *global* range of networks. Additional primary ethical consequences of this distribution cluster about a growing range of issues and duties for us as *cosmopolitans* or citizens of the world (cf. Ess, 2009, pp. 104–132).

Secondly, however, the term digital can be misleading. To begin with, analogue media and experience have not disappeared. On the contrary: Most of our "digital" devices begin and end with analogue interfaces and experiences—for example, the voice spoken into a microphone and heard through an earpiece, pictures produced on screen or on paper, the analogue clock or watch face powered by digital movement, and so forth. The term digital thus risks privileging one aspect of our technologies while neglecting other experientially significant and thus ethically crucial details and dimensions of our lives as *embodied* beings. A third difficulty here is *technological determinism*—the view that once a technology is unleashed, its further development and directions are inevitable and largely beyond the capacity of human beings to control or steer. This view is now well refuted, but to use the term digital in digital ethics may inspire a determinism that is both false and thus counterproductive.

A fourth feature of digital technologies further complicates our efforts to develop a working definition of the term digital ethics. To state the obvious: These technologies continue to develop at a dizzying, if not simply explosive pace—and this in both sheer *quantitative* as well as *qualitative* directions. *Quantitatively,* for example, hardware, software, and network access became less expensive and thus more widely accessed in a growing diversity of ways. The upshot is that the sheer scope and range of ethical issues evoked by digital media likewise continue to expand. Simultaneously, this ongoing development further entails *qualitative* changes in the role of digital technologies in our lives and thus the ethical challenges they bring in their train.

For many of us, the most notable such changes over the past 5 years accompany the increasing *mobility* of digital devices (e.g., in the form of smartphones, tablets, GPS-equipped digital cameras, etc.), along with their *locative* features, that is, their capacity to determine, record, and transmit the location of their owners. These affordances allow for a whole new range of uses, from "friend-finding" applications to "augmented reality" experiences, for example, as smartphone apps can highlight nearby services, guide us through particular tourist experiences, and so forth (Bechmann, Waade, & Ess, 2012). But mobility also means that we bring (and/or are coerced into bringing) such devices into spaces and moments of our lives that were "digital-free zones" but a few years ago. Especially as networked through the Internet and the Web, these devices and our diverse uses of them not only extend the range and scope but also the complexity of ethical challenges affiliated with digital media. This is apparent in the first instance with regard to the novel challenges to privacy such devices evoke—and that both for their users and for researchers attempting to examine such uses. Most fittingly, the first section of this volume is defined by attention to privacy issues and research ethics.

Moreover, such devices have radically transformed our contemporary understanding of journalism and the role of "ordinary citizens" in the gathering and distribution of news—most obviously, in the form of eyewitness photos and videos, for example, of police violence against demonstrators, made on smartphones, and then uploaded to the Web for a global audience. At their best, such citizen contributions to the traditional work of journalism can greatly enhance one of journalism's primary tasks, namely, fostering democratic dialogue and processes (Ward, 2011). On the other hand, if such contributions are to work against a strong tendency, especially in market-driven media outlets, toward news as infotainment—both professionals and amateurs require a new awareness and ethical sensibilities as to what sorts of content are appropriate and useful to capture and upload for the rest of the world. In short, what was a professional ethics for a comparatively limited number of professional journalists and their supporting institutions is quickly morphing into a hybrid ethics, including important dimensions of "ethics for the rest of us," insofar as our ownership and use of a smartphone and other mobile locative media render us all into potential journalists. One of the primary contributions of several chapters collected in this volume is directed precisely toward the further development of such an ethics (e.g., Roberts & Steiner, Grabowski & Yeng, as well as Dennen).

"Digital" in the term "digital ethics" thus represents a moving target, one that is constantly evolving and expanding at a breathtaking rate. But however daunting these difficulties may be, as these first examples of contributions from this volume demonstrate, the good news is that ongoing work from a variety of disciplines and ethical perspectives has issued not only an extensive literature on diverse ethical issues evoked by digital media. In addition, these contributions point toward what appears to be a reasonably stable taxonomy of signature issues, prominent approaches, and disciplines that demarcate the terrains and boundaries of digital ethics.

Digital Ethics: Past and Present

We can begin by way of the taxonomy of issues sketched out in *Digital Media Ethics* (Ess, 2009). This taxonomy emerged in light of my intention that the volume serve as an "ethics for the rest of us," that is, for more or less all human beings who pick up a digital device and are thereby confronted with a range of ethical concerns. And as we have already begun to see, this list of core issues broadly coheres with the topics and challenges addressed within this volume.

The taxonomy begins with privacy and extends through copyright, pornography, violence in video games, and global citizenship. Regarding *privacy:* in 2007 through 2009, there was much concern, for example, over the apparent willingness of young people to offer up information on social networking sites (SNSs) such as MySpace and Facebook in ways that seemed to their elders to constitute a dangerous abandonment of privacy. *Plus ça change plus c'est pareil.* At the time of this writing, Facebook now claims more than 800 million users and its various efforts to respond to those users' concerns and protests over privacy continue to be, shall we say, problematic. For its part, Google is doing no better: its most recent changes in its privacy policies, in particular, have been charged with violating European Union data privacy protections (BBC News, 2012). At the same time, however, there are striking developments in protecting *individual* privacy in online environments in one domain of the world that, until the past few decades, has upheld state authority over the individual in dramatically nondemocratic fashion—namely, north Asia, including Taiwan and mainland China (Greenleaf, 2011; Sui, 2011). (We return to this apparent paradox in the final section on selfhood.)

Privacy concerns, moreover, are not solely of the moment for individuals as citizens and users of digital devices: They are of first importance for researchers from more or less every discipline who focus on the various human engagements and interactions facilitated by digital media. Not surprisingly, the ethics of research on our online engagements has likewise grown remarkably over the past decade or so, beginning with the development and publication of the ethical guidelines for online research by the Association of Internet Researchers (AoIR; Ess & Jones, 2002). To be sure, a host of new ethical challenges for researchers have emerged alongside more recent developments, as referenced by the notion of "Web 2.0," one highlighting new forms of engagements such as blogs and microblogs, social networking sites, sites such as YouTube for posting "prosumer" material, and so forth. Ethical reflection has followed suit, including the ongoing work of the AoIR ethics guidelines committee, which has recently developed a document complementing the 2002 guidelines (Markham & Buchanan, 2012). From these perspectives, it is especially fitting that this volume begins with an extensive section titled "Research Ethics." Chapters included in this first section address several of the central and compelling issues concerning privacy and research ethics and constitute a significant contribution to both the literatures on privacy and Internet research ethics (cf. Buchanan, 2011).

By the same token, the debates over copyright and copy left (including Creative Commons licensing frameworks) have become more complex and developed over the past 5 years. It is not simply that technological developments continue to make copying and sharing copyrighted materials ever easier and faster, despite ongoing efforts by various industry organizations to dampen such behavior. Perhaps most notably, at least in the West, the emergence of Pirate Bay as a political party—that is, one that articulates an explicit philosophy and political agenda, with growing presence and legitimation in at least the Scandinavian countries—suggests a sea-change in our views on nothing less fundamental than what counts as property and who has a right to access it. In this volume, the chapter by Brian Carey titled "Permissible Piracy?" most directly takes up this thread.

And—surprise!—*pornography* has not gone away, neither as a major component of digital production, consumption, and distribution via networks, nor, as a result, as evoking a wide range of ethical concerns. The ethical issues here become (yet again) increasingly subtle and difficult, as made clear here in the chapter by Jo Ann Oravec titled "The Ethics of Sexting: Issues Involving Consent and the Production of Intimate Content." More broadly, the ethical issues surrounding pornography have also become more complex because of the multiple ways in which technological developments lead to an increasing *blurring* of once clear lines between the "actual" or the "real" (as primarily material) and the "virtual" (as grounded in diverse computational technologies). A particular example recently developed by Litska Strikwerda (2011) is especially striking: What ethical judgments can we make (if any) regarding the production, distribution, and consumption of *virtual* child pornography, that is, with the equivalent of photoshopped "children" instead of their real-world counterparts? Such purely virtual material thereby avoids real-world harm to real-world children—and thus (largely) escapes the capacity of the (once) prevailing ethical frameworks of deontology and utilitarianism to ground ethical analyses and judgments. If, however, our moral intuitions are that there remains something ethically problematic about such material, these intuitions can be articulated and powerfully argued through the framework of *virtue ethics* (Strikwerda, 2011, pp. 155–158). This particular issue at the increasingly blurring boundaries between real and virtual not only demonstrates how much more complex our ethical issues and analyses have become: Strikwerda's analysis further exemplifies the growing importance of virtue ethics in our efforts to deal with such issues (Ess, in press).

At the same time, more recent research has shown, at least within the European Union that pornography and sexually explicit materials (SEMs) are not the primary issue or problem for young people, however their parents may fear it to be so. While unwanted encounters with SEMs are problematic, for young people, the more extensive and compelling problem is that of cyber bullying (Livingstone, Haddon, Görzig, & Ólafsson, 2011, p. 24).

As for *computer games* and particular concerns with violence, briefly put, the debate rages on. Every study that claims to show some sort of causal linkage between extensive engagement with violent games and real-world violent behavior evokes a hail of critique and counterexamples that point out, for example, various weaknesses in the methodology, and so forth. But in other ways, ethical reflection and discussion of the issues evoked by computer games have dramatically expanded and developed, in part as our focus on the ethical dimensions of computer games now extends well beyond the once-dominant concern with violence. The exemplar here is Miguel Sicart's work, *The Ethics of Computer Games* (2009). Very happily, this volume includes Miguel's chapter, "Instrumental Play or the Moral Risks of Gamification," as well as a chapter titled, "Griefing Through the Virtual World: The Moral Status of Griefing," by Roland Wojak, which focuses on a distinctive ethical issue within MMORPGs (Massively Multiplayer Online Role Playing Games).

Finally, the ethical issues of *global citizenship* have, likewise, rapidly expanded in both scope and depth. My primary focus in 2007 to 2009 was on the cross-cultural encounters—and, often, conflicts—that the Internet and the Web made increasingly commonplace. These days, once again, it seems that the difficulties and issues have only become more complex. On the one hand, the problems of cross-cultural encounters remain: The iconic event of these encounters, the publication of the Mohammed cartoons by *Jyllands Posten* in Denmark in 2006, and the resulting aftermath, still echoes in contemporary Danish life as well as in journalism studies. But in the meantime, our digital lives have led to ever-greater surveillance of our activities, and this in a range of forms, beginning with what Anders Albrechtslund (2008) described as "participatory" or "voluntary" surveillance, that is, our horizontal attention to one another precisely through SNSs, microblogs such as Twitter, and so forth. There are certainly salutary effects and benefits of such engagements. But more darkly, the "digital shadow"—see Erin Reilly in this volume—that results as part of the digital records of our online lives allows for ever-more sophisticated forms of data-mining, for example, driven first of all by commercial interests in commodifying us as consumers and markets (Livingstone, 2011; Smythe, 1994). Simultaneously, especially since the terrorist attacks of September 11, 2001, the nation-states of the West have justified increasing scrutiny, including suspension of due process rights, in the name of protecting citizens from such attacks. Such surveillance is increasingly enhanced by a growing web of biometric technologies—that is, not simply increasingly ubiquitous surveillance cameras, but such cameras as feeding, say, facial- and gait-recognition software that allows specific individuals to be quickly identified.

Once again, these technological developments appear to threaten first of all what have been foundational Western conceptions of individual privacy as a positive good and as a basic human right. As we are about to see, these conceptions of privacy are closely tied to even more foundational assumptions regarding our senses of selfhood and identity, and thereby, moral agency. These assumptions in turn play a large role in our determining which ethical frameworks are most legitimate and most appropriate to take up in our analyses and efforts to resolve ethical issues. In addition to the sorts of expansions and transformations of DE that were discussed earlier with regard to what we might think of as the substance of DE, that is, the range and complexity of issues within its purview, there are correlative developments in these more foundational areas as well those that increasingly command our attention.

Digital Ethics: Presents and Futures

As I see these things, the further development of DE will first of all continue along the trajectories we have seen here: Presumably, ongoing technological development and diffusion of digital devices will continue to drive a parallel expansion in the range, scope, and complexity of ethical challenges evoked by these devices. At the same time, the more foundational components of DE—specifically, our ethical frameworks and our basic assumptions regarding selfhood and agency—will also undergo transformations and developments.

For example, we can briefly examine *robot ethics* as an example of a substantive development and extension of DE, one that will surely become of ever-greater significance. Once the provenance of science fiction, robot ethics has quickly emerged as a field in its own right, driven primarily by sometimes astonishing advances in robotics. Three primary areas stand out here: the use of robots in healthcare and eldercare (including so-called care-bots); multiple uses of robots in combat, including "warrior-bots" of increasing sophistication and growing roles in real-world warfare; and, of course, "sex-bots." Such devices certainly include an array of digital components—various computational devices, including artificial intelligence (AI) and "Autonomous Agents" (AAs). These devices also often rely on Internet-enabled communication, along with other digital technologies such as GPS systems. Correlatively, there is a rapidly expanding, interdisciplinary development of robot ethics. In particular, some robotic devices are sufficiently autonomous as to be capable of choosing between at least a narrow range of possible behaviors—including, in the case of warrior-bots, whether a potential target is a "legal" target as defined by Just War theory, or an innocent/noncombatant who is not a legitimate target for lethal force (cf. Wallach & Allen, 2010, p. 73ff.). Manifestly, there are a host of ethical issues in play here, many of which are novel, insofar as we human beings appear to be increasingly "offloading" both the labor and moral responsibilities of caring and warfare to such autonomous devices.

Hence we should include robot ethics within the field of DE. Not only will robots become increasingly part of our lives in the developed world, as these first examples suggest, as Sherry Turkle has now famously argued, their doing so may well threaten our own ethical and related human sensibilities and capacities (2011). The worst-case scenarios include not simply our increasing inability to develop and marshal the capacities and virtues required for deep friendships, long-term familial relationships, and so forth, but also the very possibility of democracy itself (a theme we return to shortly).

Such developments, moreover, point to still more foundational issues and challenges concerning our senses of selfhood and ethical frameworks. To begin with, as I have argued elsewhere (Ess, 2010), Medium Theory (as developed by Harold Innis, Elizabeth Eisenstein, Marshall McLuhan, and Walter Ong, and elaborated more recently by the work of Naomi Baron [2008], among others), a range of empirical findings from Internet studies, and then recent changes in privacy laws, points to fundamental changes in our conceptions of selfhood, identity, and moral agency in both Western and Eastern cultures and traditions. To begin with, Medium Theory shows strong correlations between our modalities of communication (*orality, literacy, print,* and then the "secondary orality" of electric media, including the communications we enjoy via networked digital media), our notions of selfhood and identity, and thereby our conceptions of proper or legitimate social structures and political regimes. In these terms, the primary orality of preagricultural peoples correlates with a *relational* (and strongly *emotive*) sense of selfhood, that is, one largely defined precisely by its relationships with others; such relational selves, moreover, accept both strongly hierarchical social structures and nondemocratic political regimes. The rise of literacy begins the development of a novel conception of the self—what emerges in the modern West in conjunction with the extension of literacy in part by way of the new technologies of printing. What becomes articulated as a primarily *rational* and *individual* conception of selfhood then correlates with both democratic regimes and greater emphasis on equality, including (eventually) gender equality. In Ong's (1988) terms, "electric media," including digital media, reintroduce many of the characteristics of *orality* into our communicative landscapes, a "secondary orality." This can be seen especially in the new communicative possibilities of Web 2.0, which expands the use of the auditory as well as the visual. Moreover, such secondary orality brings with it a (re)turn to the relational and the emotive as primary features of selfhood. For example, social networking sites are precisely about articulating and expanding the relationships defining oneself, and the available "like" buttons, coupled with limited capacities for text (i.e., 140 characters in any given tweet), appear to favor the emotive over the complexly rational.

That our sense of selfhood is changing in these ways—i.e., from more emphasis on the self as a rational autonomy to a greater emphasis on the relational and emotive self—is further consistent with Western shifts in attitudes away from both individual privacy as an *exclusive* right toward "group privacy" or "publicly" as the blurring of the boundary between individual privacy and public spaces (e.g., David, 2009).By the same token, the turn against individual property as an exclusive right, in favor of property as shared or inclusive, is manifest in a host of ways, including the rise of Pirate Bay as a political party arguing precisely against earlier notions of property as protected by copyright. What is striking in these shifts, finally, is that such *inclusive* notions of privacy and property are in fact primary features of oral, preagricultural societies, societies also marked by hierarchy and nondemocratic values.

The correlations highlighted by Medium Theory thus force the question: Will such (more) relational-emotive selves, as favoring more inclusive notions of privacy and property, likewise turn away from modern Western notions of democracy and equality and become more comfortable with more hierarchical and nondemocratic regimes? The development of privacy laws over the past decade alone suggests some interesting answers. As we have seen, both the United States and the European Union have reduced individual privacy protections, including due process rights. By contrast, North Asia has *increased* individual privacy protections in law (if not always in practice), including discussion of due process rights, even in mainland China (Greenleaf, 2011; Sui, 2011).

In addition to these large political implications, these transformations of selfhood will inevitably entail shifts in our understandings of what ethical frameworks are most legitimate and appropriate for use, both within DE and beyond. We have already begun to see such transformations: As noted earlier, the case of virtual child pornography highlights the rising importance of *virtue ethics* in DE. More broadly, in fact, virtue ethics emerged originally among societies marked precisely by more *relational* senses of selfhood: It may be no accident, that is, that as our sense of selfhood in the West increasingly emphasizes relationality; we find ourselves more and more interested in virtue ethics at large. This means, finally, that the ethical frameworks of deontology and utilitarianism that have predominated in the modern West for at least 2 centuries are now being called into question. In this volume, similarly radical questions are raised in the contribution from Anthony Beavers: "Could and Should the Ought Disappear from Ethics?"

Concluding Remarks

I hope this overview has made clear that DE is a field that is rapidly growing in both scope and depth, thereby constituting an array of issues whose complexity thus expands quasi-exponentially and in multiple directions simultaneously. This is especially true, not only in terms of a growing range of ethical topics forced upon us by a growing diversity and use of digital devices, as illustrated here by way of the example of robot ethics. At the same time, because digital media appear to force transformations in the most fundamental components of any ethics—namely, conceptions of selfhood and thereby of ethical frameworks *per se*—DE must come to grips with these deeply philosophical and complex issues. In this direction, DE will incorporate more and more of what is now the provenance of Information and Computing Ethics, as well as of philosophy proper (Ess, in press)—meaning that it will be more and more difficult to sustain a DE focusing on ethics as "ethics for the rest of us."

Nonetheless, more or less everybody who takes up a digital device will thereby be confronted with the ethical issues and challenges evoked by the ever-growing, ever-more complex technological ecology that increasingly shapes and defines our lives. We have no choice but to pursue DE as an "ethics for the rest of us." I hope it is also clear, however, that the chapters constituting this volume contribute in no small ways to DE as such an ongoing enterprise. They thereby offer us both substantive insight and encouragement for continuing our work.

References

Albrechtslund, A. (2008).Online social networking as participatory surveillance. Retrieved from http://www.uic.edu/htbin/cgiwrap/bin/ojs/index.php/fm/article/view/2142/1949

Baron, N. (2008). *Always on: Language in an online and mobile world.* Oxford, UK: Oxford University Press.

BBC News. (2012, March 1). Google privacy changes "in breach of EU law." Retrieved from http://www.bbc.co.uk/news/technology-17205754

Bechmann, A., Waade, A.M., & Ess., C. (2012). *Locative mobile media: Mobile internet, cross-media, place and performativity.* Manuscript in preparation.

Buchanan, E. (2011). Internet research ethics: Past, present, and future. In M. Consalvo & C. Ess (Eds.), *The handbook of Internet studies* (pp. 83–108). Oxford, UK: Wiley-Blackwell.

David, G. (2009). Clarifying the mysteries of an exposed intimacy: Another intimate representation mise-en-scéne. In J. C. Kristóf Nyíri (Ed.), *Engagement and exposure: Mobile commu-*

nication and the ethics of social networking (pp. 77–86). Vienna, Austria: Passagen Verlag.
Ess, C. (2009). *Digital media ethics.* Oxford, UK: Polity Press.
Ess, C. (2010). The embodied self in a digital age: Possibilities, risks, and prospects for a pluralistic (democratic/liberal) future? *Nordicom Information, 32*(2), 105–118.
Ess, C. (in press a).Introduction to special issue, "Who am I Online?" *Philosophy and Technology.*
Ess., C. (in press b). Ethics at the boundaries of the virtual. In M. Grimshaw (Ed.), *The Oxford handbook of virtuality.* Oxford, UK: Oxford University Press.
Ess, C., & Jones, S. (2002). Ethical decision-making and Internet research: Recommendations from the AoIR ethics working committee. Retrieved from http://aoir.org/reports/ethics.pdf
Greenleaf, G. (2011) Asia-Pacific data privacy: 2011, year of revolution? *UNSW Law Research Paper No. 2011–29.*Retrieved from http://papers.ssrn.com/sol3/papers.cfm?abstract_id =1914212
Livingstone, S. (2011). Internet, children, and youth. In M. Consalvo & C. Ess (Eds.),*The handbook of internet studies* (pp. 348–368). Oxford, UK: Wiley-Blackwell.
Livingstone, S., Haddon, L., Görzig, A., & Ólafsson, K. (with members of the EU Kids Online Network).(2011). EU Kids Online: Final report. Retrieved from http://eprints.lse.ac.uk /24372/
Markham, A., & Buchanan, E. (2012). Ethical Decision-Making and Internet Research (version 2.0): Recommendations from the AoIR Ethics Working Committee. Retrieved from http://aoirethics.ijire.net/
Moor, J. (1985). What is computer ethics? *Metaphilosophy,16*(4), 266–275.
Ong, W. (1988). *Orality and literacy: The technologizing of the word.* London, UK: Routledge.
Sicart, M. (2009). *The ethics of computer games.* Cambridge, MA: MIT Press.
Smythe, D. W. (1994). Communications: Blindspot of western Marxism. In T. Guback (Ed.), *Counterclockwise: Perspectives on communication* (pp. 266–291). Boulder, CO: Westview Press.
Strikwerda, L. (2011). Virtual child pornography: Why images do harm from a moral perspective. In C. Ess & M. Thorseth (Eds.), *Trust and virtual worlds: Contemporary perspectives* (pp. 139–161). Oxford, UK: Peter Lang.
Sui, S. (2011, October).*The law and regulation on privacy in China.* Paper presented at the Rising Pan European and International Awareness of Biometrics and Security Ethics (RISE) conference, Beijing, China.
Turkle, S. (2011). *Alone together: Why we expect more from technology and less from each other.* Boston, MA: Basic Books.
Wallach, W., & Allen, C. (2010). *Moral machines: Teaching robots right from wrong.* Oxford, UK: Oxford University Press.
Ward, S. (2011). *Ethics and the media: An introduction,* Cambridge, UK: Cambridge University Press.

Section 1

Research Ethics

Introduction by Meghan Dougherty

This section presents three distinct approaches to the ethical conduct of research in the digital age and provides provocative discussion framing each approach. Wyatt broadens the discussion by questioning standard guidelines of study design by offering a holistic approach that addresses the impact of research beyond the researcher-respondent relationship. Dennen delves deep, revealing the complexity behind the public/private dichotomy that often guides digital research ethics. Carpenter and Dittrich challenge the definition and application of the "human subjects research" label used by U.S. institutional review boards (IRBs) to aim for more practical and altruistic standards for computer science research that often crosses over into Internet research and social science territories. Behind all of these approaches lies the fundamental question of our relationship with technology and how these technologies are shaping and reshaping the academic research environment.

Wyatt addresses this fundamental question head on by discussing the concept of e-research. She defines this growing field within a changing research environment, evolving from a strong history of ethical consideration. This changing environment poses new ethical challenges. To offer advice on identifying and navigating the complex ethical questions faced in e-research study design, Wyatt uses a simplified version of the research process as a heuristic device and offers not only advice on frequently occurring ethical questions in research but also provides a structure for evaluating unanticipated ethical concerns. Wyatt argues that we need to broaden our discussion of e-research ethics beyond the simple definitions—those of researcher-respondent relationships, which proscribe respondent privacy and anonymity as paramount—that most often guide ethical decision making in study design. Wyatt provides a succinct definition of e-research and places it in a rich history of evolving ethics in research methods. She argues that a broader vision should be used to guide ethical decision making in study design and offers a structured heuristic for identifying ethical conundrums in research and making the right decision to proceed.

Where Wyatt argues for a broader consideration of research ethics beyond research-respondent relationships and protecting the privacy and anonymity of the respondent, Dennen dives deep into the important question of respondent privacy expectations. Dennen begins with the all-too-often oversimplified research paradigm that treats all content as data. Dennen recognizes the force of this data-driven approach in research and aims to pause long enough to unpack the repercussions of simply assuming that public content online is permissible data for research. Dennen argues that a person's sharing of content publicly online does not necessitate his or her permission to be included in scholarly research. To illustrate this point, Dennen discusses the act of publishing online and the meanings that may be derived from such an activity. Given such online acts, different general privacy issues can develop. Dennen discusses the possible expectations of privacy a respondent may have given any number of publishing acts online and the different ways researchers assuming that public content equals permissible data can impact respondents unknowingly.

Dennen's chapter provides examples for consideration when designing a study in digital culture following U.S. IRB guidelines. A number of scenarios are identified in which further consideration may be warranted in order to more inclusively assess the impact of researchers' approaches to online public content as data. Dennen leaves the reader with more questions than answers, illustrating the point that privacy online is a multifaceted concept in need of more complex consideration than existing public/private dichotomy that can guide study design.

Carpenter and Dittrich approach digital research ethics from an alternative perspective—one that is often overlooked in favor of the ones identified in the other chapters in this section. *Bridging the Distance* addresses complex and interconnected systems—specifically in computer security research—that tie together an intricate mix of stakeholders whose risk of harm may be a confusing point in determining what kind of research falls under the rubric of "human subjects" and so requires IRB oversight. The authors explain that the nature of computer security research puts a distance between researchers and potentially impacted parties, causing these researchers to believe that they are not interacting with the humans that their study impacts and so do not consider their research to be "human subjects research" in line with IRB definitions. Regardless of this research perspective, the authors point out that, "[t]here are not consistent ethical standards for considering and measuring the adverse effects of computer science research on human subjects or society." The computer security systems we build, and the subsequent findings from research on those systems, have an impact on human lives given how inextricably bound computer systems are with everyday life.

Carpenter and Dittrich review IRB standards as applied to computer security research and focus specifically on informed consent standards. They point out that in such research, anyone with an Internet connection may be potentially harmed by the research activities or findings. Given current definitions, and the quickly broadening scope and complexity of this research, there is a fundamental gap between informed consent treatments called for by IRB guidelines and the reality of what is practical and feasible in terms of informing potential participants of risks. Carpenter and Dittrich unpack key definitions and guidelines from IRB standards and federal regulations, apply them to computer science study design practices, and suggest that new definitions and standards are needed on both sides of the equation.

As technology, and our relationship with it, evolves, so too will the academic research environment. We will continue to use new tools to conduct research and study evolving behaviors enabled by new tools, and so we will continue to rethink our ethical standards for research. The authors in this section recognize the importance of reevaluating ethical standards for research as methods for studying and objects of study evolve in a shifting technological landscape. In these chapters, our authors challenge current standards. They offer suggestions for how to build an ethical study in evolving technological landscapes. And they offer direction for how to reconsider the application of existing standards to evolving technological landscapes that may adequately address the conceptual gaps about ethical issues between the people who use tools for everyday living and the people who use them for research.

CHAPTER ONE

Ethics of e-Research in Social Sciences and Humanities

Sally Wyatt

Since their earliest days, computers and networked computers have been used in academic, government, and commercial research settings (Abbate, 1999; Agar, 2006). Computers are extremely good at storing and processing structured data. Powerful tools can help researchers to analyze and represent complex data. Since what we now call the Internet went public and commercial in the early 1990s (Thomas & Wyatt, 1999), the digitally mediated communication of ordinary people has proved to be a fascinating source of data, only intensified with the rise of Web 2.0 or social media. The use of transactional data generated by billions of people going about their everyday activities is of enormous value not only to market researchers and intelligence services but also to scholars in the humanities and social sciences (Savage & Burrows, 2007). The digitization of library and archival holdings can immensely facilitate some of the laborious processes of scholarly work.

The sometimes real and sometimes only apparent novelty and ubiquity of digital tools and digitized data bring questions of research ethics into relief. It is those questions, and some answers, which are the focus of this chapter. But first, the concept of e-research itself is situated within the context of a changing academic research environment. I then provide an overview of the history of research ethics, before examining a range of ethical issues raised by the growth of e-research, using a simplified version of the research process as a heuristic device. The final section provides some guidelines for conducting e-research ethically, arguing that we need to move beyond a simple focus on the researcher-respondent relationship and the accompanying focus on the privacy and anonymity of respondents in order to encompass a broader range of both relationships and research processes. In order to do this, I suggest that it is important to acknowledge the range of different ways that digital technologies affect scholarly research.

The Changing Research Context

Long-established hierarchies and practices of scholarly knowledge production are undergoing change, challenged from within and by broader social developments. Four more or less generic developments can be distinguished, namely *growth, accountability, network effects, and technology.*[1] The first two are primarily associated with the decline of autonomous science in practice and as an ideal and instead point to the myriad ways in which a wider group of social actors is involved not only in setting the research agenda but also in knowledge creation itself. The third and fourth processes relate to the sociomaterial relations and conditions of knowledge production.

First, the growth of the university system after World War II in advanced industrialized countries was part of an overall growth in research and development activities. It was accompanied by an increase in the diversity of students, staff, and subjects. Universities became accessible to a wider range of people. Many new (inter)disciplines emerged, such as media and cultural studies and women's, gender, and queer studies. Sometimes, as in the case of women's studies, these new disciplines could be attributed to the greater diversity of people entering universities. On other occasions, the emergence of a new field is related to the widespread availability of a new object or medium such as television in the case of media studies.

Second, researchers are increasingly called to account for the quality and quantity of their output and its relevance for nonacademic social actors. National and international science and other policy makers have played an increasingly important role in steering and evaluating academic output in both teaching and research. A wider range of social actors, including for-profit corporations (Radder, 2010) as well as civil society organizations (Epstein, 1996), are no longer simply the passive recipients of knowledge produced elsewhere but are increasingly active in its production. While many positive outcomes can be observed arising from new forms of governance and participation, modes of accountability nevertheless have a range of effects, positive and negative, on the work of scholars and on institutions (Strathern, 2000).

Third, the apparent success of "big" science in the physical and biological sciences in the postwar period means that large teams working across institutional and disciplinary boundaries have become the ideal for research managers and policy makers. Science policy makers have been engaged in promoting particular forms of work organization by investing in research infrastructures and by encouraging large-scale collaboration often on an interdisciplinary and/or international basis (Bell, Hey, & Szalay, 2009; Hine, 2008). Collaborative networks of researchers, across disciplines and/or countries, have emerged as a new ideal organizational form (Castells, 1996). Moreover, a research field known as network science has emerged that systematically analyzes network effects in nature and society (Scharnhorst, Börner, & van den Besselaar, 2012).

Last, but by no means least for this chapter, digital technologies have been taken up in all stages of knowledge production (Dutton & Jeffreys, 2010; Hine, 2005; Jankowski, 2009; Wouters, Beaulieu, Scharnhorst, & Wyatt, in press), and they play an important role in the processes mentioned earlier. Digital technologies shape scholarly collaboration, monitoring, and evaluation of academic output as well as the communication of results both within the academic community as well as to wider audiences.

I focus on the social sciences and humanities (SSH), though the developments outlined earlier affect all disciplines, and, as I suggest later, the ethical practices adapted within the SSH are themselves shaped by those used in the life sciences. Given that society and the products of human creativity and interaction are the main objects of study in SSH, the implications of changing modes of governance and organization are likely to be different for SSH than for the STEM (Science, Technology, Engineering, Medicine) disciplines. Changes in society not only reconfigure the conditions under which new knowledge is produced; they are at the same time the object of study, what Giddens (1984) called the double hermeneutic. These changes take place concurrently and interactively, though not always in the same direction or at the same rate. They cannot simply be reduced to developments in research technologies (which is certainly the view of some policy actors). However, focusing on the use of technology provides a valuable analytical focus, giving indications about changing circumstances of use and about the way traditional hierarchies, locations, and processes of knowledge production in the SSH may be disrupted. What can be seen from history is that the Internet has been "pushed" as a medium for promoting efficiency gains in the delivery of education, the organization of universities, and increasingly in the conduct of research itself, all in the context of a set of policies for the "information society." Close examination of this discourse reveals the central role given to technologies as both cause and effect of the information society. Thus, information and communication technologies are understood as determining this new stage of postindustrial development (the cause) whilst, at the same time, university managers, research councils, and governments argue that the push to deliver education online or to develop computational methods is merely a rational response to the coming of the information society (the effect), reflecting a "justificatory" form of technological determinism (Wyatt, 2008).

There are a number of analytical advantages to taking digital technologies as a starting point for understanding the ethics of SSH research. It is not the technical tools *per se* that are most interesting but the ways in which digital tools stimulate reflection about objects, methods, and practices of research, including ethical practices. For example, most efforts at using digital technologies in research involve the formalization of research practices and of research objects. Such processes require actors to reflect on their practices and negotiations about standards for communication or for data formats. These processes make explicit aspects of research and values of research communities, which themselves provide an interesting starting point for analysis as well as for interdisciplinary collaboration. Such negotiations may enhance the reflexivity of practitioners and the likelihood of being able to enter into discussion with them about their work (Beaulieu & Estalella, 2011).

A number of concepts are already in circulation, such as Mode 2 (Nowotny, Scott, & Gibbons, 2001), postnormal science (Funtowicz & Ravetz, 1993), technoscience (Haraway, 1985; Latour, 1987), and the triple helix (Leydesdorff & Etzkowitz, 1998), which aim to capture the way in which knowledge production has been changing over recent decades. There are also more specific concepts which draw attention to the ways in which digital technologies have affected the research process used by actors in the field (such as e-science, cyber infrastructure, grid, semantic web) as well as by academic colleagues (including digital humanities, computational humanities, collaboratories, virtual research environments). Each of these terms has a particular connotation, and none arrives free of history and context. Some of these histories have been recounted recently. (See, e.g., Jankowski (2009) for *e-science* and *e-research,* Borgman (2007) for *cyber infrastructure* and *digital scholarship,* Hine (2008) for *cyber science,* and Wouters et al. (in press) for *virtual knowledge.*)

The term *e-research* (Jankowski, 2009; Wouters & Beaulieu, 2006; Wouters et al., 2008) is used here because it evokes an interesting object of study, and because it foregrounds the practice of research rather than hardware or infrastructure. It acknowledges those many forms of research that are not reliant on high-performance computing, thereby including the use of new media and digital networks. Thus, *e-research* indicates an object of research and intervention that is more inclusive of a variety of research modes. Sensitivity to disciplinary practices (Fry, 2006; Kling & McKim, 2000) is therefore a key assumption in the use of *e-research.* The evocative nature of e-research also serves to catalyze hopes, resistance, and controversies. Controversies have a long tradition within science and technology studies (STS) and history of science as moments that are analytically rich (Martin & Richards, 1995). They have been also identified as "early warning indicators" for paradigmatic struggle, new ideas, methodological innovations, and ethical reflections, the subject of the next section.

New Ethical Considerations

The changing landscape in which academic researchers find themselves, as briefly sketched earlier, has implications for the ways in which research ethics are understood and practiced. The institutionalization of ethical review is, at least in part, a response to the developments affecting universities more generally, described earlier as "growth," "accountability," "network effects," and "technology." The first two are related, in that growth has contributed to the perceived need for formal accounting mechanisms, accounting not only for the outputs of research but also for the ways in which it is conducted. As research systems become larger and more specialised, institutional and disciplinary norms are codified. Of course, the declining trust in science by the public (European Commission, 2007, aka *Wynne Report*) has also played a role in strengthening ethical review. These changes are further complicated by the growth of international and interdisciplinary research teams, where national and disciplinary norms may not be easily reconciled. New and emerging science and technologies themselves present all sorts of ethical challenges, especially in the life sciences, where developments often raise profound challenges to basic notions of what it is to be human.

Many new techno-scientific developments are investigated from a broader variety of disciplines, including social sciences, law, and philosophy, often going under the label of ELSI (Ethical, Legal, and Social Issues). Collaboration among these disciplines is not completely without friction. ELSI is often used to address questions related to genetics, nanotechnology, animals (in research), and food. In many countries, ELSI research is implemented in the form of accompanying programs to major research initiatives, particularly in the fields of genomics and nanotechnology. This is meant to ensure early reflection on the possible impacts such research might create on the societal level. ELSI research is taking on the role that would have been occupied by technology assessment research until recently, putting normative and regulatory questions center stage.

This changing landscape is important for understanding ethics, but scientific research ethics has its own history and dynamics. Based on their analysis of key documentary sources and interviews with a range of relevant social actors in the United States, Montgomery and Oliver (2009, pp. 141–147) identified three institutional logics guiding research conduct. These are associated with particular time periods, though the earlier logics are still visible in later time periods. The first, described as "science in the pursuit of truth," 1945–1975, is idealized in the Mertonian (Merton, 1942) norms of universalism, communalism, disinterestedness, and organized scepticism. Scientists were assumed to be socialized into these norms through their education and training. Science was effectively self-regulated and did not need formal codes, neither at an institutional nor disciplinary level. The second, "preventing scientific misconduct," 1975–1990, was largely prompted by growing public awareness of questionable research practices which exploited people by exposing them to risk, such as the Tuskegee Syphilis Study (Retherby, 2009) and the Milgram experiments (Herrera, 2001). Reports of publicly funded fraudulent research were on the rise. A series of initiatives in the United States by different government bodies, including the National Institutes of Health and Congress itself, began to impose stricter requirements for universities wanting to receive public funding. It was in this climate that local IRBs (Institutional Review Boards) were established to approve all research protocols and consent procedures for research involving human subjects. More formal definitions of scientific misconduct were put forward, and universities were obliged to develop policies for responding to allegations of misconduct. The third stage, "promoting research integrity," from 1990 onward, was based on the realization that conducting responsible research was more complex than simply not harming people and avoiding plagiarism and falsification of data. Montgomery and Oliver also identified new threats to the realization of Mertonian norms, arising from the changing research environment, especially the increased role of private funding (Mode 2 science as mentioned earlier). In particular, disinterestedness and universalism are challenged by research funding regimes which aim to valorize the results of research for private gain, through patents or commercial secrets.

As already mentioned, these three logics can coexist, and the dynamics vary between countries. The institutionalization of ethical review and scientific misconduct is less strong in European countries, where attempts to introduce ethical review are often met with scepticism or more active resistance (Hedgecoe, 2008). While ethical review for medical and psychological research is well established, it is only in the past decade that review is being introduced for SSH research. This is partly related to the "return of the subject," following critiques of poststructuralism and postmodernism, which underestimated the integrity of the subject. Based on an ethnographic analysis of how UK ethical review committees dealt with social science (itself part of a four-country comparative study), Hedgecoe (2008) urged his social science colleagues to stop their special pleading to be exempt from such procedures on the grounds that interpretative social science work necessarily requires openness and flexibility in the course of research. He pointed out that sociologists and anthropologists are perfectly capable of explaining this to funding bodies when necessary, so it should also be possible to explain this to ethical review committees. Furthermore, his observation of the work of UK review committees was that they were not *a priori* unsympathetic to interpretative sociology. Hedgecoe emphasized that research designs may also evolve in bio-medical research and that all are concerned with treating people with respect and protecting them from harm, including any harm that might arise from embarrassing exposure. Opening up one's research design to colleagues should only improve the process and not be seen as censorship or restriction.

Ethics of e-Research

What do these dynamic research and ethics landscapes mean for doing e-research, not only research that takes the digital as its object but also research that is done differently, in which digital tools and methods are intimately bound up with the practice of research? In this section, I focus on the debate about what to do with the information found online, often freely provided by individuals. Can researchers simply see this as a windfall, or do they have ethical obligations to those who placed it there? I also look at some of the ethical and normative aspects of "openness" in the publication of research results and of data. But the use of digital tools and methods may also give rise to ethical reflection at all stages of the research process. These are summarized in Table 1.

The most discussed ethical questions facing researchers is what to do with the vast quantities of data online, especially data generated by people engaging in everyday online activities. Is this data simply to be treated as any other publicly available information? If so, then appropriate citation and acknowledgment of sources solve many problems. For much information found online, that is indeed more than adequate. However, in informal settings, including games, fan sites, and patient groups, such an approach would not meet the basic requirements outlined earlier, of protecting people, sometimes from themselves.

These debates, around questions of the privacy and anonymity of respondents, have been well-rehearsed (among others, Ess, 2009; Ess & AoIR Ethics Working Committee, 2002). Problems arise because protecting the human subject is seen as the primary obligation of both researchers and ethical review committees. It is assumed there is a simple, clear relationship between data found online and individuals in the real world. This is what Carusi (2008) called "thin" identity, and what Beaulieu and Estalella (2011) referred to as "traceability." Given that individuals may reveal a great deal of personal information online, researchers have to be concerned to protect the anonymity and privacy of research subjects. The danger facing social

Table 1. Ethical Questions Associated with Use of Digital Technologies in Research

STAGE IN RESEARCH PROCESS	DIGITAL TECHNOLOGY: APPLICATION OR TOOL	ETHICAL QUESTION	OTHER STAKEHOLDERS
Literature review	Search engines and databases	What are the effects of (secret) search engine algorithms on the availability of information?	Libraries
Identifying participants	Search engines, social networking sites	When is it acceptable for researchers to use online resources to gather information about respondents? What happens when respondents use online resources to find out about researchers? Should one be "friends" with respondents?	Respondents
Data collection	Multiplayer games, Web 2.0 sites, online forums	When is lurking an acceptable research strategy? How can anonymity and privacy of respondents be protected?	Respondents
Data analysis	Software for managing data, data mining, computational tools and methods	Should researchers rely on the possibilities for analyzing textual data in a quantitative manner? What are the implications of the formalization of data?	Software designers

Table 1. (*continued*)

Data sharing	Distributed databases, virtual research environments, open access publications	What are the implications for categorization (categories always involve inclusion and exclusion)?	Other researchers, publishers
Data storage and preservation	Cheap storage media for digital information, annotation, and metadata tools	Should all (publicly funded) research data be stored? Under what conditions can or should data be reused?	Other researchers, funding agencies, archives, and libraries
Data representation	Visualizations	Do certain kinds of information become more or less visible when using advanced visualization tools?	Readers, publishers
Authorship and acknowledgment	Authoring software, distributed databases, enhanced publications	How can the inputs of technical support be properly acknowledged? Are participants in online environments to be treated as "respondents" or as "authors"?	Research support staff, technical support, respondents

researchers and their respondents is that individuals could be easily identified and traced if, for example, their words, avatars, or nicknames are mentioned in academic texts. Even when attempts are made to anonymize individuals, other details could more or less inadvertently enable others to identify groups or individuals. For example, Zimmer (2010) analyzed the "Tastes, Ties and Time (T3)" research project based on the Facebook accounts of a cohort of university students. He pointed out that despite measures taken to protect the anonymity of the students, it did not take too long to identify the university and even individual students, based on analysis of the course offerings and other demographic data.

Individuals going about their everyday online lives are not obliged to be part of academic research. Even if they are voluntarily providing extensive details about their innermost thoughts or about what they ate for breakfast on their blogs or tweets, it does not necessarily mean that this information is fair game for researchers. It is not always adequate for researchers to say this information is in the public domain. Nissenbaum (2010) discussed this in terms of "contextual integrity" and drew attention to the expectations of privacy people may have in particular contexts, online and offline. She highlighted the right to privacy "neither as a right to secrecy nor a right to control but a right to appropriate flow of personal information" (Nissenbaum, 2010, p. 127). Similarly, Bakardjieva and Feenberg (2000) pointed to the dangers of alienation arising from indiscriminate use of material found online, alienation experienced by people who have provided information in one context who may understandably not be happy to find it taken up in another.

Carusi (2008) and White (2002) made a different argument regarding the use of data found online. They both pointed to the dangers associated with assuming an isomorphic relation between an individual and some online utterances. They both argued for treating online material as forms of representation. By doing so, other ethical issues and responsibilities emerge. For White (2002), it is important to recognize the constructed nature of online material, so that researchers can challenge the abundance of hate speech that is easily found. She argued that by recognizing the highly mediated and representational character of online material and by considering the ethical codes of literary studies or of art history and visual culture, different sorts of research questions could be addressed, and she pointed out that this would open up different forms of analysis. Carusi (2008) also suggested that the focus on the privacy of research subjects restricts the range of ethical questions that can and need to be addressed with the rise of e-research within the humanities and social sciences. Bishop (2009) also bemoaned the fact that a focus on the privacy of respondents makes it more difficult to consider the ethical obligations researchers may have to other actors and stakeholders. I now turn to perhaps less obvious moments in research when the use of digital technologies throws up ethical questions, starting with literature reviews and then moving into publication of results and of data.

As students, we learn that research begins with a review of the literature. Nowadays, that task has become much easier with the use of search engines and databases. But this may already be more ethically complex than it seems, not least because of our increasing reliance on proprietary search engines and databases. Using different computers with different settings, or even the same computer at different times, may yield different results. As Introna and Nissenbaum (2000) argued more than a decade ago, search engines systematically exclude certain sites and types of sites. This techno-politics of design, code, and algorithm remains invisible to most users but has implications for the results of our searches. (See also Carpenter & Dittrich, this volume, for a discussion of how ethical review committees tend to ignore computer research as being outside their purview.)

The naïf response to this would be to point to the great promise of the Internet to provide free and easy access to information, not only to scholarly articles and books but also to original data. One could argue that it does not matter where the host or website is based, so long as people all over the world can access data and information. To further this goal, in 2003, many academies, universities, research councils, and institutes adopted the *Berlin Declaration on Open Access to Knowledge in the Sciences and Humanities*. There are now more than 100 signatories, mostly from Europe but also from South, North, and Latin America. Open access is defined "as a comprehensive source of human knowledge and cultural heritage that has been approved by the scientific community" (Max Planck Society, 2003, p. 1). The declaration identifies the Internet as the most important tool for making "original scientific research results, raw data and metadata, source materials, digital representations of pictorial and graphical materials and scholarly multimedia material" (Max Planck Society, 2003, p. 1) freely available. The signatories are committed to finding ways of developing existing legal and financial frameworks to make open access possible.

There are challenges to realizing the objectives of the *Berlin Declaration*, not least the long-standing practices of scientific publishers. Many scientific journals have "article processing charges," which can be as much as U.S. $5,000. Sometimes there are additional charges simply to submit an article for consideration and for color printing. For example, the *Journal of Neuroscience* charges authors a $120 submission fee, $950 publication fee, plus $1,000 for each color figure and an optional $2,500 "open access" fee (BioMed Central, 2011). These sums are far beyond the means of many universities. Sometimes fees are automatically waived for authors based in poorer countries, but often exemptions have to be sought on a case-by-case basis. In these instances, "open access" is a new business model for publishers whereby authors pay instead of or as well as readers. This has consequences for the distribution of knowledge production, with richer disciplines and universities having greater opportunities for publishing their research results. These and other practices (Sismondo, 2009) bring into serious question scientific principles of transparency, disinterestedness, and peer review.

Charging authors for publication is rare in the SSH, not least because such departments are usually less well funded than their natural science counterparts, even within a single university. But charging practices can cause problems for those in the SSH who study ethical, legal, and social issues relating to science and technology and who wish to communicate their results to a natural science audience. There are other important differences between the disciplines. One of the aims of the *Berlin Declaration*, as mentioned earlier, is that there should also be greater access to data. Much of this discussion assumes a computational view of what science and research are about. In other words, data are collected, and then, in the interests of openness, digitally deposited and preserved so that others can use those data to replicate the results of others and to test new hypotheses. But scholars in the interpretative humanities and social sciences work with different kinds of data, in which the context of data collection is integral to its interpretation and understanding. Defining species of plants or insects is already difficult; coming to agreement on occupational codes in order to make comparisons about the work people do across time and country even more so. Making sense of qualitative interview data about, for example, people's understanding of health and illness, collected by someone else, is almost impossible.

Moreover, there are reasons why open access for data and data sharing may be resisted, especially by smaller and less powerful researchers and research groups. There are few incentives for sharing data within the research system and even fewer for doing the difficult and time-consuming work needed to ensure that data are compatible and accessible in meaningful ways. For many types of qualitative data, the privacy of research subjects and participants may be compromised by open access (Wouters et al., 2007). Some countries, such as Canada, require researchers to destroy data after 5 years, precisely in order to protect research participants. This is a different ethical principle from open access, but nonetheless, it is an important one. However, as Bishop (2009) suggested, reuse of qualitative data may not be as practically or ethically difficult as qualitative researchers often assume.

Conclusion: The Importance of Being Earnestly Ethical[2] and of Taking Technology Seriously

In this chapter, I have situated e-research within the context of a changing academic research environment. I suggested that academic research is undergoing a number of changes associated with growth, increased accountability, the success of "big science," and the use of digital technologies in all stages of knowledge production. Despite the proliferation of terms to capture the latter, I choose to use "e-research," as it opens up a wider range of moments and modes of research and does not simply reduce e-research to the application of computational methods and models. I also provided an overview of the history of research ethics, not least in order to understand why the debates about e-research have tended to focus on the privacy, confidentiality, and anonymity of respondents. I look again at this debate before also examining some other ethical issues raised by the growth of e-research, especially those associated with open access debates.

There are two important conclusions, which can be summarized in terms of being earnestly ethical and of taking technology seriously. First, being "earnestly ethical" means thinking with and beyond the formalized requirements of ethical review boards. Ethical review has been institutionalized, often for good reason. Rather than seeing this as an obstacle to be overcome, it is important to work with such committees in order to improve our research designs. Having done so is not the end of our ethical responsibility, however. We also need to move beyond a simple focus on the researcher-respondent relationship and the accompanying focus on the privacy and anonymity of respondents in order to encompass a broader range of both relationships and research processes. In order to do this, I suggest that it is important to acknowledge the range of different ways that digital technologies affect scholarly research. Doing e-research can throw up ethical and normative questions at any moment in the process and may affect researchers' relationships not only with research participants but also with colleagues, funders, and users of research.

Second, we need to take technology seriously. Technology is not a neutral tool that gives us unmediated access to data, references, ideas, and people. Technology mediates and structures researchers' interactions at all stages of the research process. While such tools offer us many possibilities and may well reduce the time and effort associated with some tasks, we should not remain ignorant of the ways such tools may render invisible some literature, information, data, categories, institutions, or people.

Acknowledgments

Many people have influenced the development of my thinking on this topic. For those whose written work has been influential, see the references. I would like to take this opportunity to thank Maria Bakardjieva, Anne Beaulieu, and Annamaria Carusi in particular, and other colleagues who participated in events we organized about the ethics of e-research under the auspices of the Virtual Knowledge Studio, 2008–2010. I would also like to thank the organizers and participants of the International Symposium on Digital Ethics, held at the Center for Digital Ethics and Policy at Loyola University Chicago, on October 28, 2011, as well as the participants at the meeting of the Graduate Research School, Faculty of Arts and Social Sciences, Maastricht University, about online research ethics, December 7, 2011. All errors of fact, value, and judgment are my own.

Notes

1. This section is adapted from Wyatt, Scharnhorst, Beaulieu, and Wouters (in press).
2. Phrase used by Maria Bakardjieva during a presentation on June 8, 2008, Amsterdam, the Netherlands.

References

Abbate, J. (1999). *Inventing the internet.* Cambridge, MA: MIT Press.

Agar, J. (2006). What difference did computers make? *Social Studies of Science, 36*(6), 869–907.

Bakardjieva, M., & Feenberg, A. (2000). Involving the virtual subject. *Ethics and Information Technology, 2,* 233–240.

Beaulieu, A., & Estalella, A. (2011). Rethinking research ethics for mediated settings. *Information, Communication & Society, 15(1),* 23–42 Retrieved from http://www.tandfonline.com/doi/abs/10.1080/1369118X.2010.535838

Bell, G., Hey, T., & Szalay, A. (2009). Computer science: Beyond the data deluge. *Science, 323*(5919), 1297–1298.

BioMed Central. (2011). *Comparison of BioMed Central's article-processing charges with those of other publishers.* Retrieved from http://www.biomedcentral.com/about/apccomparison/

Bishop, L. (2009). Ethical sharing and reuse of qualitative data. *Australian Journal of Social Issues, 44*(3), 255–272.

Borgman, C. L. (2007). *Scholarship in the digital age: Information, infrastructure and the Internet.* Cambridge, MA: MIT Press.

Carusi, A. (2008). Data as representation: Beyond anonymity in e-research ethics. *International Journal of Internet Research Ethics, 1*(1), 37–65.

Castells, M. (1996). *The information age: Economy, society and culture, Volume I: The rise of the network society.* Oxford, UK: Blackwell.

Dutton, W. H., & Jeffreys, P. W. (Eds.). (2010). *World wide research: Reshaping the sciences and humanities.* Cambridge, MA: MIT Press.

Epstein, S. (1996). *Impure science: AIDS, activism, and the politics of knowledge.* Berkeley: University of California Press.

Ess, C. (2009). *Digital media ethics.* Cambridge, UK: Polity Press.

Ess, C., & AoIR Ethics Working Committee. (2002). *Ethical decision-making and internet research: Recommendations from the AoIR ethics working committee.* Retrieved from http://aoir.org/reports/ethics.pdf]

European Commission. (2007). *EUR 22700—science & governance—taking European knowledge society seriously.* Luxembourg, UK: Office for Official Publications of the European Communities. Retrieved from http://ec.europa.eu/research/sciencesociety/document_library/pdf_06/european-knowledge-society_en.pdf

Fry, J. (2006). Studying the scholarly web: How disciplinary culture shapes online representations. *Cybermetrics: International Journal of Scientometrics, Informetrics and Bibliometrics, 10*(1), 2. Retrieved from http://cybermetrics.cindoc.csic.es/articles/v10i1p2.html

Funtowicz, S., & Ravetz, J. (1993). Science for the post-normal age. *Futures, 25,* 739–755.

Giddens, A. (1984). *The constitution of society.* Cambridge, UK: Polity Press.

Haraway, D. (1985). Manifesto for cyborgs: Science, technology and socialist feminism in the 1980s. *Socialist Review, 80,* 65–108.

Hedgecoe, A. (2008). Research ethics reviews and the sociological research relationship. *Sociology, 42*(5), 873–886.

Herrera, C. D. (2001). Ethics, deception, and "those Milgram experiments." *Journal of Applied Philosophy, 18*(3), 245–256.

Hine, C. (Ed.). (2005). *Virtual methods: Issues in social research on the Internet.* Oxford, UK: Berg.

Hine, C. (2008). *Systematics as cyberscience: Computers, change and continuity in science.* Cambridge, MA: MIT Press.

Introna, L., & Nissenbaum, H. (2000). Shaping the web: Why the politics of search engines matters. *Information Society, 16*(3), 1–17.

Jankowski, N. W. (Ed.). (2009). *E-Research: Transformation in scholarly practice.* New York, NY: Routledge.

Kling, R., & McKim, G. (2000). Not just a matter of time: Field differences and the shaping of electronic media in supporting scientific communication. *Journal of the American Society for Information Science, 51*(14), 1306–1320.

Latour, B. (1987). *Science in action.* Cambridge, MA: Harvard University Press.

Leydesdorff, L., & Etzkowitz, H. (1998). The triple helix as a model for innovation studies. *Science and Public Policy, 25*(3), 195–203.

Martin, B., & Richards, E. (1995). Scientific knowledge, controversy, and public decision making. In S. Jasanoff, G. E. Markle, J. C. Petersen, & T. Pinch (Eds.), *Handbook of science and technology studies* (pp. 506–526). Thousand Oaks, CA: Sage.

Max Planck Society (2003). *Berlin Declaration on open access to knowledge in the sciences and humanities.* Retrieved from http://oa.mpg.de/files/2010/04/berlin_declaration.pdf_

Merton, R. (1942). Science and technology in a democratic order. *Journal of Legal & Political Sociology, 1,* 115–126.

Montgomery, K., & Oliver, A. L. (2009). Shifts in guidelines for ethical scientific conduct. How public and private organizations create and change norms of research integrity. *Social Studies of Science 39*(1), 137–155.

Nissenbaum, H. (2010). *Privacy in context: Policy and the integrity of social life.* Palo Alto, CA: Stanford University Press.

Nowotny, H., Scott, P., & Gibbons, M. (2001). *Re-thinking science: Knowledge and the public in an age of uncertainty.* Cambridge, UK: Polity Press.

Radder, H. (Ed.). (2010). *The commodification of academic research. Science and the modern university.* Pittsburgh, PA: University of Pittsburgh Press.

Retherby, S. M. (2009). *Examining Tuskegee. The infamous syphilis study and its legacy.* Chapel Hill: University of North Carolina Press.

Savage, M., & Burrows, R. (2007). The coming crisis of empirical sociology. *Sociology, 41*(5), 885–899.

Scharnhorst, A., Börner, K., & van den Besselaar, P. (Eds.). (2012). *Models of science dynamics—encounters between complexity theory and information sciences.* Berlin, Germany: Springer.

Sismondo, S. (2009). Ghosts in the machine: Publication planning in the medical sciences. *Social Studies of Science, 39*(2), 171–198.

Strathern, M. (2000). *Audit cultures: Anthropological studies in accountability, ethics and the academy.* London, UK: Routledge.

Thomas, G., & Wyatt, S. (1999). Shaping cyberspace: Interpreting and transforming the Internet. *Research Policy, 28,* 681–698.

White, M. (2002). Representations or people? *Ethics and Information Technology, 4*(3), 249–266.

Wouters, P., & Beaulieu, A. (2006). Imagining e-science beyond computation. In C. Hine (Ed.), *New infrastructures for knowledge production: Understanding e-science* (pp. 48–70). Hershey, PA: Idea Group.

Wouters, P., Beaulieu, A., Scharnhorst, A., & Wyatt, S. (Eds.). (in press). *Virtual knowledge.* Cambridge, MA: MIT Press.

Wouters, P., Hine, C., Foot, K. A., Schneider, S. M., Arunachalam, S., & Sharif, R. (2007). Promise and practice of open access to e-science. In W. Shrum, K. R. Benson, W. E. Bijker, & K. Brunnstein (Eds.), *Past, present, and future of research in the information society: Reflections at the occasion of the World Summit on the Information Society, Tunis, 2005* (pp. 159–172). New York, NY: Springer.

Wouters, P., Vann, K., Scharnhorst, A., Ratto, M., Hellsten, I., Fry, J., & Beaulieu, A. (2008). Messy shapes of knowledge—STS explores informatization, new media, and academic work. In E. Hackett, O. Amsterdamska, M. Lynch, & J. Wajcman (Eds.), *Handbook of science and technology studies* (pp. 319–350). Cambridge, MA: MIT Press.

Wyatt, S. (2008). Technological determinism is dead. Long live technological determinism. In E. Hackett, O. Amsterdamska, M. Lynch, & J. Wajcman (Eds.), *Handbook of science and technology studies* (pp. 165–180). Cambridge, MA: MIT Press.

Wyatt, S., Scharnhorst, A., Beaulieu, A., & Wouters, P. (in press). Introduction: Virtual knowledge. In P. Wouters, A. Beaulieu, A. Scharnhorst, & S. Wyatt (Eds.), *Virtual knowledge.* Cambridge, MA: MIT Press.

Zimmer, M. (2010). "But the data is already public": On the ethics of research in Facebook. *Ethics and Information Technology, 12*(4), 313–325.

CHAPTER TWO

When Public Words Are Not Data

Online Authorship, Consent, and Reasonable Expectations of Privacy

VANESSA P. DENNEN

Researchers do not need consent to write about the public acts of celebrities or politicians, but do high traffic Internet-based authors or commenters count as famous people? And, is anyone who posts to a public blog, YouTube channel, or Twitter account creating data that a researcher can use freely? What is the researcher's responsibility to the people who generate his or her data? These types of questions guide decisions about when, how, and with what permissions publicly available Internet-based data sources may be used.

Merely referring to words and images on the Internet as "data sources" depersonalizes them for the researcher, separating them from their authors. However, those authors, no matter how publicly they share their words or images, may still desire privacy in some contexts. They share their words, images, and often their lives online for myriad personal and professional reasons but not so that they can become the objects of scholarly research. They do not think of themselves as participants until they find out that a researcher has considered them as such.

Published research findings can and often do affect participants. When participants either read firsthand or learn secondhand (and sometimes inaccurately) of a researcher's published findings, the effect may take various forms, including an emotional toll or a change in the person's relationships or ability to trust others (see Brettell, 1996, for examples). Further, Internet-based research based on unique phenomena or that includes direct quotes may be readily traced back to the originating source. Even paraphrases or loosely anonymized descriptions might be sufficient for Internet sleuths to identify an original source. Additionally, research reports bring public behavior that may be limited to one arena or setting to another arena or setting, exposing people who may have felt a certain false security writing under pseudonyms or writing on low-traffic websites. The cumulative effect of these issues is that researchers in Internet-based settings may have an unintended negative impact on the people they study.

This chapter addresses ethical issues related to conducting research using Internet-based sources such as blogs, Twitter, and discussion forums, with a focus on whether the people posting to these forums should have a reasonable expectation of being asked to consent before their online words and images are collected and analyzed by researchers and steps researchers should take to help protect the privacy and identities of these people. It begins with a brief discussion of publishing online, followed by an overview of privacy issues as they relate to Internet-based communication in general. Finally, it addresses issues related to assessing a participant's desire for privacy, handling consent, and researcher presence, and managing the impact that data presentation and reporting may have on participants.

Throughout this chapter, I make reference to research in a university setting with an IRB serving as a gatekeeper for the research process. This context is admittedly U.S.-centric and represents the system of research that I know best, but the larger ethical issues that I raise are ones that impact researchers regardless of country, institutional affiliation, and presence of an ethical oversight board or committee. Additionally, I draw upon reports of published research and my own experiences as a researcher to demonstrate different approaches that have been taken in Internet-based research. My intent is simply to have examples for discussion, not to render judgment on any particular researcher's behavior.

Publishing Online and Public Figures

The Internet has created a more democratic form of publishing than traditional mass media, and, as a consequence, a new type of celebrity or public figure has emerged. There are no gatekeepers standing at the door deciding who can appear on a computer screen. If you want to share your life online, you may. All you need is access to a computer and the Internet. Any number of sites will host your words and images for free. Bloggers such as Heather Armstrong (http://dooce.com) and Jenny Lawson (http://theblogess.com), Twitter users such as Justin Halpern (http://twitter.com/shitmydadsays), and video series such as lonelygirl15 (http://www.youtube.com/user/lonelygirl15; now known to be fiction, but initially considered a video diary) have attracted large numbers of followers and mass media attention. Heather Armstrong has incorporated her blog into her career, making money from ads, appearances, and books related to her blog, among other things. Jenny Lawson writes for a number of media outlets. Justin Halpern's Twitter activities have spawned a book and a television series. These individuals consciously place themselves on the public stage and seek, or at least accept, celebrity. Informed consent would not be needed to analyze their online writings any more than an author's permission would be needed to analyze his book or a news anchor's permission would be needed to analyze his delivery style.

Next, consider your average mommy blogger, knitting blogger, or lifestyle blogger with a public profile, a lengthy blogroll, and an active group of commenters. These bloggers outnumber the Heather Armstrongs and Jenny Lawsons. They may withhold details such as real name, location, and profession out of safety or privacy concerns, and it may be unclear if they are striving toward some level of celebrity. Do these bloggers fit in the same category for research purposes? Should a researcher have an ethical responsibility to request permission before using a blog as a data source?

The questions I ask here push beyond the one addressed in Eastham's (2011) discussion of privacy, research design, and informed consent for blog-based studies. I am concerned not only with what online data researchers may use without informed consent to the satisfaction of an IRB but also with the larger ethical issue of *when* it is acceptable for us to conduct research based on other peoples' online activities—and in particular, their words and images—without their consent or even their knowledge. Further, I am concerned with our ethical responsibility as researchers to consider the potential impact of our research on our participants, their online practices, and their relationships with both other people online and the people who they write about in an online forum.

Perspectives on Online Privacy

Levels of Privacy

Online privacy sounds like an oxymoron, but some online spaces are more private than others. As people have increasingly engaged in online activities, they have traded privacy for the ability to be social in new ways (Papacharissi & Gibson, 2011). This assertion is supported by studies such as Ploderer and colleagues' (Ploderer, Howard, & Thomas, 2010) research on amateur and celebrity participation in bodybuilding forums, which found that public profiles and open access forums played an important role in building the community of users, particularly among amateurs.

Peoples' expectations of privacy vary according to online medium. When sending personal e-mails, an individual might expect the message contents to remain private, although a hacker or a recipient who elects to share the message with a wider audience could violate that privacy. At the other end of the spectrum is a totally open forum. For example, someone commenting at http://www.cnn.com is contributing words to a very public and widely marketed virtual space.

Sveningsson Elm (2009) recommended viewing public and private as end points on a continuum. There are various points between totally closed and totally open online communication systems, and with each system comes a different level of access challenge for the researcher as well as expectation of privacy for the user. A class discussion in Blackboard is password protected, and contributors have a reasonable expectation that their instructor serves as the gatekeeper; most students would anticipate only class members to be reading their contributions. We might consider such communication to be semiprivate under Sveningsson Elm's system.

Other communication may be semipublic, meaning access is restricted in some manner (e.g., requires a log-in), but access or membership is available to all who desire it. To contribute to or read a discussion on the forums of the *Tallahassee Democrat* (http://www.tallahassee.com) newspaper website requires a paid subscription and user profile. And anyone can set up a Facebook account, but they need to do that before they can post to a "public" Facebook group.

Even fully public blogs may enact some level of privacy restrictions in terms of posting and commenting while leaving access to others open. The administrative owners of blogging accounts determine who may post to a blog, and they may require that commenters be logged in to the blogging system as well, all while maintaining a public face. Thus, password-protected systems can represent a variety of expectations about who might be reading and discussing at each site.

Assessing Privacy

The concept of privacy in an online medium must be examined from multiple angles in a research perspective. Relevant questions include the following: What is the intent of the author? What are the restrictions on data access? How personal or revealing are the data? How will the data be reported? Essentially, the researcher needs a way to assess privacy.

Password-protected data sources require access and consent and generally carry the designation "private." So, when bloggers use password protection and pseudonyms, or when they exclude their blogs from search engine indexing, a researcher can assume that these actions indicate a desire for privacy (Eastham, 2011). However, the ephemeral qualities of otherwise public data—nonarchived and editable sources in particular—may lead some individuals to feel their actions are less public than they really are (Buchanan, 2011). Coupled with use of pseudonyms and user names, participants' resulting perceptions of anonymity online may lead them to engage in greater self-disclosure than they otherwise might be comfortable with (Hookaway, 2008). Additionally, Internet use is not a neatly compartmentalized activity for most people; instead, its use occurs in the midst of our other life activities (Bakardjieva, 2011). The end result is that people freely share about their offline worlds in online settings.

Assessing the desire for privacy in online spaces, however, is a far more complicated issue than looking at tool and archive settings. For the individual, there is a trade-off between the ability to have privacy and the ability to build social capital in an online environment (Ellison, Vitak, Steinfeld, Gray, & Lampe, 2011). Electing to communicate publicly does not mean that privacy is not desired. Various studies (e.g., Dennen, 2009; Orgad, 2005) have highlighted instances in which people discuss private issues in public spaces, often taking some precautions to obscure identity, because the desire to seek support or interact with others online somewhat outweighs (but does not eradicate) the opposing desire for privacy. In my own research, participants often have surprised me when commenting about their identity and privacy concerns; those whom I might expect to be quite protective of their words and identity often are not and vice versa. Thus, external technical factors alone are insufficient for assessing a person's comfort with a researcher analyzing and sharing his or her words. Requesting consent to participate and direct interaction, then, are the only reliable ways to assess a participant's desire for privacy.

Shifting Perspectives from Researcher to Subject

Qualitative research practices typically acknowledge the impossibility of objectivity, with a tradition of qualitative researchers seeking to learn and triangulate the participants' perspectives. During fieldwork, qualitative researchers frequently become directly involved with their participants and their practices. Techniques such as prolonged engagement and participant observation facilitate the researcher's ability to learn the participant perspective, whereas other techniques such as bracketing assist with maintaining a separate sense of self as researcher.

However, the availability of rich qualitative data online has resulted in a great deal of archival research. Internet researchers do not always participate in the practices they are studying or engage with the people they are studying. If he or she has not interacted with participants, engaged in similar practices, or served as a research subject in this context, a researcher's understanding of the participant perspective—why the participants communicate with others online, with whom they hope to interact, and how they might feel about being researched—is naturally limited.

Papacharissi and Gibson (2011, p. 79) suggested that even when people's actions are played out on a public forum, they have a "right to make decisions about their own path to privacy, sociality, and publicity." Although their context is sharing via social networking and not research, I believe the same sentiment applies. Just because a person shares information online and a researcher can freely access it does not mean that the individual is consenting to participate in research. In this context, a researcher would be both providing commentary on the individual's shared information and publicizing it to a different audience.

Text or Personal Conversation?

One's perspective on the nature of online writing impacts how one perceives its level of privacy. Is an online diary an act of personal communication? Are the messages in a discussion forum a personal conversation? Or are these and other forms of public online writing simply texts? As a social scientist, I have always considered them to be personal communications and conversations that I might happen to overhear, much as I often hear the personal conversations of people in public spaces such as shopping malls, airplanes, and restaurants. However, in the last several years, I have had some interesting conversations with humanities scholars, who would treat a corpus of written work published online as a text, no different for them to incorporate in their scholarship than a novel or essay published in a mainstream media outlet.

Personal conversations have a sense of impermanence that traditionally published texts do not share; spoken words, if not recorded, can only be remembered and retold. Memories are fallible, and retellings are interpretations. Many forums in which individuals write online exist somewhere between personal conversations and traditionally published texts in terms of their permanence in that they are continuously editable and able to be deleted yet also potentially archived.

Personally, I find it difficult to view these online data sources as texts. For me, tweets and discussion forums represent conversations. And blogs often have a diaristic quality—not intended to be shared with a general audience, although they sometimes fall into stray hands. There is potential to cause harm to the people we study by treating their works as texts, particularly when personal information has been shared. The following blog post, quoted with the author's permission and anonymized at her request, vividly demonstrates how real people and real-life consequences lie behind what is written online:

> This will be my last post as [Pseudonym]. Last night, I learned from a family member that someone sent posts from my blog to my family of origin. These were posts where I vented and worked out some emotions that I do not share with family members,

> mainly because I don't want our relationship to consist of me screaming at them and them screaming at me.
>
> As many of you know, I have been blogging anonymously on a variety of platforms for 5.5 years. My blog has been a diary for me, a place where I vent, blow off steam, and express things I can't and don't say to anyone.
>
> I never, ever wanted any family member to read my blog, anymore than I would have wanted them to read my teenage diary. It was a place for me to work through my own issues and feelings so that I wouldn't say hurtful things to anyone later.
>
> I blogged under a pseudonym not only to protect my own privacy, but also that of my university, my friends and colleagues, and my family members. Pseudonymous blogging, especially by those of us who are academics, entails a certain type of trust that members of the community will not out each other. I am extremely hurt that someone betrayed me in such a fashion.

Although this blogger was hurt by a deliberate malicious event, not through research, she nonetheless had expected a certain degree of privacy within her public online community, only to see that trust betrayed. Her blog had been a safe place, and her sense of safety was removed. In response, her blogging practices changed.

A similar chain of events is potentially at risk of being set off when researchers call attention to public communities that desire a degree of privacy. People seek safe places online where they can relax, reflect, or find support. Violating trust by not seeking consent can harm that sense of safety and make the consent process more difficult for future researchers.

My Perspective on Privacy

What I bring to this discussion is the perspective of a researcher who has also been a participant in similar research and who has directly discussed with some of my own research participants how they feel about the types of attention they might receive from different sources when living their lives online. Through these experiences, I have developed a position on conducting Internet-based research with human participants that pushes me to continuously consider how my participants might feel if they were to read what I publish, to seek consent in some situations when I am confident that my institution's IRB would permit me to conduct the research with a waiver of consent.

I engage in public online discourse, both under my given name and pseudonyms. When I use a pseudonym I do not believe myself to be anonymous, but I am consciously and purposefully distancing my words and actions from my everyday personal and professional identities. I prefer to control if, when, and how these identities get connected.

I have knowingly been a participant in research on class discussion boards. I have been asked to participate in research as a blogger, and at times I have chosen to participate and at other times I have not. I was unwittingly included as a participant in an e-mail-based study, informed of my participation only after the fact and without the opportunity to choose about consent. And while I have participated in online discourse via Twitter and on a variety of discussion forums, I have not knowingly been included in a study.

I recognize the possibility that my blog corpus has been included in someone else's research without my knowledge, and I am not entirely comfortable with that thought. Of particular concern is that my writing might be taken out of context by someone looking for a particular sound bite or that items intended as isolated postings, written over a long time span and across different sites, might be assembled in a manner that encourages misinterpretation. Essentially, I want to feel that I have some degree of control over my words even though I know that once they are shared in a public place I have lost that control. I simply maintain hope that a researcher—particularly one doing a qualitative study—would invite my involvement as a participant to provide contextual perspective and that she would also demonstrate a strong desire to respect my own level of comfort about how the data are used and presented. These are, perhaps, naïve hopes, but I try to be mindful of them when researching other peoples' online activities.

Privacy and Consent Issues

The ability to access large datasets from one's own computer without interacting with people can be attractive, and obtaining consent from online participants can be a complex process that at times is simply not practicable. Contacting everyone in a large Twitter-based dataset could be an unwieldy process. Some bloggers do not include an e-mail address on their blogs. Commenters in some forums may post as "anonymous," and some forums may not provide access to user profiles or e-mails or provide an internal private messaging system. Waivers of consent are often possible when data are online and publicly available, allowing research that is invisible to the participant to take place. IRBs will frequently grant a waiver of consent, especially for large, archived datasets with hundreds of contributors that are going to be used in an aggregate analysis. Note, however, that a waiver of consent does not guarantee that participants, should they find out about the study, will not be upset (see Merz, 2010, for a story about an Internet-based deception study with waiver of consent that upset some participants).

The Institutional Review Board's Role

Although an IRB is effectively a gatekeeper for faculty research at universities, the decisions made by these boards are not always as consistent or as informed as a researcher might hope. The IRB's task is to ensure that research is planned and executed in such a way to minimize harm to participants and to disallow studies in which the anticipated harm outweighs the benefits. However, IRBs deal with diverse types of research, varying widely in discipline, method, and data sources. The members of an IRB may not be sufficiently familiar with Internet-based tools to grasp the potential privacy concerns from either a technological or interpersonal perspective. Although professional organizations such as the Association of Internet Researchers make available guidelines for Internet-based research (Ess & The AoIR Ethics Working Group, 2002), these guidelines are not necessarily reflected in the approval decisions made by an IRB. Although I believe strongly that IRBs take their role in the research compliance process very seriously and strive to minimize harm to participants at all times, I think their task is a challenging one that leaves room for the possibility of an error in judgment based on insufficient explanation of a study by a researcher or insufficient understanding of the medium, its uses, and its affordances by a reviewer.

In my own experiences with IRBs, I have found differences from institution to institution and even from reviewer to reviewer within a particular institution. For studies that do not need to be approved by the full board at a meeting, individual reviewers give feedback. I have had the experience of submitting two studies for review that were essentially identical and receiving quite different feedback from the two reviewers; in the end, both studies were approved, but one was more liberal in its interpretation of when waiver of consent might apply to the participants who had contributed to the dataset. Other times I have received feedback which suggested that my reviewer was not at all familiar with the type of data I planned to work with or the access, privacy, and consent issues that might be involved.

As an avid reader of both a number of research-oriented listservs and articles that comment on IRB treatment of Internet-based research, I know that my experiences are not unique. Zimmer (2010) noted that the Harvard University IRB failed to see the ways in which anonymity could be compromised in the Tastes, Ties, and Times dataset. Thus, it is incumbent on the ethical researcher to not rely on the IRB as the sole determinant of acceptable standards but instead to introduce their IRBs to the ethical standards and guidelines being developed by their professional associations and to help their IRBs develop internal standards and policies that can readily accommodate new technologies and the research that accompanies them.

"Invisible" Studies and the Need for Consent

From the researcher's perspective, a distinction can be made between studies that involve some form of intervention or interaction with people and those that simply work from existing texts (Eastham, 2011). IRBs require that consent be obtained—barring approval of a waiver—whenever a direct researcher-participant intervention or interaction, such as an experiment, interview, or face-to-face observation, will occur.

Intervention- and interaction-free studies involving publicly available data can be entirely invisible to the participants who provide those data. However, the lack of interaction between researcher and participant during the data collection process does not mean that informed consent is unnecessary. The sensitivity of the information being collected also should be taken into account when determining the need for informed consent (McKee & Porter, 2009). Individuals retweeting links to top technology news stories may provide relatively little personal information, and the topics of their tweets typically are not controversial. However, people who seek knowledge about and support for chronic health conditions on discussion forums tend to disclose personal information and discuss sensitive topics. And there is a gray zone in between, with the myriad online outlets in which people discuss topics ranging from politics to parenting.

Even when considering where a data source lies in the dual continuums of public/private and sensitive/nonsensitive described by McKee and Porter (2009), the final determination of whether informed consent is necessary—or, if not legally required, ethically appropriate—can be quite challenging. At first glance, a blogger who posts film reviews may be categorized as public and nonsensitive. A blogger writing about addiction may be categorized as public and sensitive. However, the film reviewer might be publishing personal notes in a public manner, with the intent of sharing them with a small group of familiar people, whereas the addiction blogger might be cultivating a reputation as a public speaker about addiction. Thus, not everyone considers so-called sensitive topics to be private, and some individuals might feel discomfort having their thoughts on seemingly innocuous and impersonal topics analyzed and disseminated in another forum.

Finally, research may not always be as invisible as we think. The bloggers whom I have studied are quite aware of the public/private balance they desire online, and many monitor server statistics. Some are quite unnerved when they note someone systematically reviewing their blog archives or a sudden flurry of activity from a new IP address or referring URL. Their reactions have included acts such as editing posts, removing archives, and, in extreme cases, deleting their blogs. They do not necessarily mind being researched, but they do want to know who combs through their archives and why.

Making Research Visible

Making online research visible requires direct researcher action. As a participant observer in an online forum, Kendall (2002) realized that her virtual presence and note-taking activities were not obvious to her participants. She initially provided regular reminders of her presence and activities, but with time she assumed that forum participants were aware. Ploderer et al. (2010) sought permission from interview participants and created a profile on the bodybuilding forum they were studying that made explicit their presence as researchers. That said, in their article it is unclear if they directly sought permission to use photos posted on the forum, which are published with blacked-out eyes. Although the authors present data suggesting that forum participants post their photos to seek a large audience, when those photos appear in a journal amidst an analysis of the bodybuilders' practices, the intent has changed.

My own experiences researching bloggers are similar to Kendall's experiences. If I wanted to use their words or write about them as individuals, I contacted them to request consent, making them aware of my presence as both a researcher and as a participant observer. Unlike these other researchers, I chose to not create a researcher profile. That decision was made for three reasons. First, a blog-based community is more dispersed than a discussion forum or online world. To have people notice my presence and profile I would have to deliberately post messages and call attention to that profile, which, in a blog-based community, would be challenging and likely ineffective at reaching participants. Second, knowing how important membership is in this community and some bloggers' distaste for "drive-by" commenters, I risked being categorized as such if I presented myself in that way. Third, comments do not garner the same visibility as top-level posts.

Because I took a personal contact approach to making my research visible, I did not include blogs by people who were unaware of my study, even though those blogs were available as a data source and helped form my impressions of the larger community and culture. Further, I only cited comments from people who were aware of my presence and status as a researcher and from whom I had directly sought permission. The only exception was in the case of individuals posting as "Anonymous" who could not be located.

The process of making one's presence as researcher known to participants must be handled carefully. Hudson and Bruckman (2005) conducted a study in public chat rooms to test the effects of different ways of announcing the researcher's presence and providing consent options, including both opt-in and opt-out consent. Their findings showed the challenges of gaining consent in a public forum. In fact, their efforts to announce presence and gain consent were alternately ignored or met with hostility from the potential participants. Researchers like Kendall and I have had different experiences in part due to the use of ethnographic methods and prolonged engagement within the community. Establishing a presence as not just a researcher but also a trusted person is important when seeking consent and increasingly so as the topic of conversation among participants becomes more personal.

Privacy and Reporting Data

There is no way to be certain that a participant will not at some point in time be upset about being a research subject other than to not do the research. Typically, IRBs allow researchers to proceed with their studies so long as the benefits of conducting the research outweigh the risks to participants. That said, determining the true benefits and risks can be an uncertain process. However, there are various actions that a researcher might take to help mitigate any perception of harm or discomfort that a research participant might experience. None is perfect, and their appropriateness is entirely dependent on factors such as the aims of the study, the content area of the study, the number of participants, and the size and nature of the data set.

Anonymizing Public Data

Anonymizing research participants in published reports may be considered "an ethical norm," although it is not always a possible or desired action (Tilley & Woodthorpe, 2011, p. 199). Anonymizing data can be a challenging process, particularly if the researcher is doing qualitative work and wishes to quote from the data source. Search engines make it rather easy to find any site that is indexed, and the default position for most online publishing tools with public accounts is to allow indexing. Further complicating this issue, RSS feeds and sites like http://archive.org can carry older or deleted versions of online data for an indefinite period.

For a qualitative researcher, trained to anonymize names and other identifiers to protect the privacy of participants but at the same time to provide thick description and make liberal use of direct quotes from data sources to promote trustworthiness among a reader audience, figuring out how to best deal with anonymizing public online data can be a real challenge. The direct quotes that are desirable can provide a direct trail back to the participant, and one should never underestimate the powers of a dedicated Internet sleuth.

The controversial Tastes, Ties, and Time dataset provides an excellent example of how there really are no guarantees where anonymity of data is concerned. This dataset consists of about 1,700 Facebook user profiles collected at Harvard University. It was originally released to the public after names—including university name—were removed. In an article reporting early findings using the dataset, the researchers invited other researchers to access and make use of this "anonymized" dataset (Lewis, Kaufman, Gonzalez, Wimmer, & Christakis, 2008). The data were collected without the permission of the Facebook users, and the data collection procedures may have resulted in inadvertent inclusion of private profiles. Although the original authors likely did not mean any harm and intended to have an anonymized public dataset, other researchers quickly identified the university and class year represented in the data (for a fuller version of this story, see Perry, 2011).

Given other information that was not stripped out of the Tastes, Ties, and Time dataset, in some instances individual students were identifiable because of a unique combination of demographic attributes (e.g., race, gender, and major). This latter part is more troublesome than the identification of the university from which the dataset was collected, and, as an example, it demonstrates how difficult it can be to produce a truly anonymized dataset while still including demographic information. Zimmer (2010) provided a thorough discussion of the ways in which this attempt to anonymize the dataset fell short, concluding, among other things, that this dataset violates the rights of the individuals whose data were mined because it is an unauthorized secondary use of their data.

However, anonymizing data is not always desirable. As Tilley and Woodthorpe (2011) noted, sometimes the conditions of funding may prohibit or hamper the ability to publish anonymized accounts, and other times participants will specifically request to go on record as themselves. In my research, studying people who have independently composed online identities that they feel sufficiently represent themselves to the public world while providing a measure of privacy (Dennen, 2009), I asked participants if they would rather be represented via their existing pseudonym or via a new one. In almost every case, the participant wished to retain his or her public pseudonym rather than have a new one. Those who desired new pseudonyms either wanted to separate their comments in an interview from their postings on a blog or felt that their pseudonym no longer sufficiently represented them, but they had no concerns with having their blog quoted with their existing pseudonym.

Reporting in Aggregate

Reporting data in aggregate is one potential solution to the consent problem. When no identifiers are involved and individuals are merely presented as part of a larger trend, not only should readers of the final report be unable to identify participants, but the participants themselves should not in any way be able to pinpoint themselves within the report. For aggregate reporting to protect privacy, the dataset needs to be of sufficient size, and care must be taken to ensure that no individual or characteristic stands out in a manner that allows identification. Again, this area was a downfall of the Tastes, Ties, and Times dataset (Zimmer, 2010).

Hammill and Stein's (2011) study of D/deaf bloggers showed how, with a small sample size, aggregate reporting may not provide sufficient privacy protections. They gave a comprehensive description of how the blogs were selected, narrowing down from 372 active blogs in a specified feed to 8 blogs and 9 bloggers that met their study criteria, each of which was included in the study. The authors further noted that they considered the blogs to be public data because the bloggers knowingly published them on the Internet. Following the lead of another study (Subrahmanyam, Garcia, Harsono, Li, & Lipana, 2009), they did not seek consent from or notify the bloggers. Although blogs and bloggers are not named, the combination of specific selection criteria, small sample, and inclusion of all bloggers meeting the criteria means that it would likely be easy for a person with good Internet search skills to identify these participants. In contrast, Lynch's (2010) study of online dieting communities was sufficiently vague about the selected sample. Although Lynch fully described the participants, potentially identifying details were omitted.

Impacts of Publishing

Introducing a New Audience

When a researcher publishes findings, participants are introduced to a new audience. It would be shortsighted and naïve to think that when they posted words or images online, the participants envisioned sharing those words with a researcher, seeing them published in a journal, or hearing them spoken at a conference. At most, they might expect to be retweeted, receive some "likes," or get a link to their blog.

Given a natural human propensity for vanity googling (Madden & Smith, 2010) and the preponderance of published research that ends up online, it's not surprising that research participants are likely to access, read, and even link to and comment on studies in which they were involved. Kouper (2010) engaged in a content analysis of 11 science blogs, naming the actual blogs and providing the URLs in her manuscript. Although she did not specify why she made their identities known in the manuscript, it seems likely that it was her focus on the bloggers as an example of public engagement with science. In this case, participants and others in their blogging community read and reacted to her work, resulting in several open, blog-based discussions, some highly critical, of her article (see Zivkovic, 2010, for a blog post discussing Kouper's article and links to others).

In the blogging community that I have studied, I have been quite aware of how some bloggers experience discomfort when high-traffic websites outside their blogging community link to one of their posts. Evidence of this discomfort includes deleted posts, edited posts, and subsequent posts directed at the high-traffic website and/or their readers. Additionally, in interviews, some bloggers have noted that these links bring the risk of a wider, unsympathetic audience and make them feel less safe on their blog. It is clear that these people are managing the tension between privacy and sociability that Papacharissi and Gibson (2011) described and that they may also have realized that their sense of anonymity, which led to self-disclosure as described by Hookaway (2008), is not infallible.

Relating this phenomenon back to my research, I oppose exposing bloggers to new audiences without first seeking their consent and ensuring that they are aware of the potential consequence. For the academic bloggers in my work, this means knowing that colleagues who do not read blogs may come across some mention of their blog in a journal or a conference presentation. The risk may be small because most of these bloggers are in disciplines far afield, and I suspect their colleagues would have little interest in my work if they did not already read blogs, but nonetheless it exists.

Impact on Secondhand Participants

Another consideration is the impact that inclusion of a data source in a study might have on secondhand or indirect participants. In an online context, this means that we need to be concerned not only about protecting our direct participants—the bloggers, the tweeters, the YouTubers—but also the individuals they feature via their posts, tweets, and videos. Justin Halpern, tweeting at shitmydadsays (http://twitter.com/shitmydadsays), shares statements by his father. Halpern's activity has spawned a book and a television series based entirely on his father's sayings. This example shows how one person's words and actions—typically the person we would identify as the participant and from whom we would seek consent—can push another person's life into the limelight.

Halpern may seem like the extreme case—a Twitter feed dedicated solely to repeating another person's words—but when telling their own stories and seeking support in digital spaces, individuals typically mention the people in their lives. Blogs often provide us with life narratives of the authors, albeit fragmented ones (Walker Rettberg, 2008); they give us insight into the identities, thoughts, and events surrounding a life. The personal mommy bloggers studied by Morrison (2011) discuss their spouses and children on their blogs. Academic bloggers mention colleagues, students, friends, and family (Dennen & Pashnyak, 2008). Orgad (2005) noted how participants in breast cancer discussion forums naturally mention the people in their lives, ranging from loved ones to medical providers. Each of these examples involves third parties, and ultimately the researcher is responsible for mitigating study-related discomfort or harm for these people as well as for the direct participants.

Conclusion

In closing, privacy expectations in public online settings are complex and multifaceted. Although the people who post online in public forums do so of their own free will, the very nature of the research activity changes the context. Seeking support and socialization online is quite different from marketing oneself online, and cultivating a specific reader audience in an online forum is not a de facto indication of consent to participate in research.

Researchers can empower participants by communicating with them, ensuring they understand both how their words and images may be used and what new audiences they might attract. A few may decline consent, but others will likely welcome the attention so long as they feel a modicum of control over their data and how it is used. Some participants may change their practices after interacting with a researcher due to a new or heightened awareness of privacy concerns, but such actions could be considered a positive benefit of the research on the participants. Ultimately, researchers should seek a balance of trustworthiness that makes them proud to have both participants and peers read and comment on their final analysis, for that is an indicator that the researcher has succeeded in his or her quest to conduct an ethically sound study.

References

Bakardjieva, M. (2011). The Internet in everyday life: Exploring the tenets and contributions of diverse approaches. In M. Consalvo & C. Ess (Eds.), *The handbook of Internet studies* (pp. 59–82). Hoboken, NJ: Wiley-Blackwell.

Brettell, C. B. (1996). *When they read what we write: The politics of ethnography.* Westport, CT: Bergin & Garvey.

Buchanan, E. A. (2011). Internet research ethics: Past, present, and future. In M. Consalvo & C. Ess (Eds.), *The handbook of Internet studies* (pp. 83–108): Wiley-Blackwell.

Dennen, V. P. (2009). Constructing academic alter-egos: Identity issues in a blog-based community. *Identity in the Information Society, 2*(1), 23–38.

Dennen, V. P., & Pashnyak, T. G. (2008). Finding community in the comments: The role of reader and blogger responses in a weblog community of practice. *International Journal of Web-Based Communities, 4*(3), 272–283.

Eastham, L. A. (2011). Research using blogs for data: Public documents or private musings? *Research in Nursing & Health, 34*(4), 353–361.

Ellison, N. B., Vitak, J., Steinfeld, C., Gray, R., & Lampe, C. (2011). Negotiating privacy concerns and social capital needs in a social media environment. In S. Trepte & L. Reinecke (Eds.), *Privacy online: Perspectives on privacy and self-disclosure in the social web* (pp. 19–32). New York, NY: Springer-Verlag.

Ess, C., & The AoIR Ethics Working Group. (2002). Ethical decision-making and Internet research: Recommendations from the AoIR Ethics Working Committee. Retrieved from http://aoir.org/reports/ethics.pdf

Hamill, A. C., & Stein, C. H. (2011). Culture and empowerment in the Deaf community: An analysis of internet weblogs. *Journal of Community & Applied Social Psychology, 21*(5), 388–406.

Hookaway, N. (2008). "Entering the blosophere": Some strategies for using blogs in social research. *Qualitative Research, 8*(1), 91–113.

Hudson, J. M., & Bruckman, A. (2005). Using empirical data to reason about Internet research ethics. In H. Gellersen, K. Schmidt, M. Beaudouin-Lafon & W. Mackay (Eds.), *ECSCW 2005: Proceedings of the Ninth European Conference on Computer Supported Cooperative Work, 18–22 September 2005, Paris, France* (pp. 287–306). Dordrecht, The Netherlands: Springer-Verlag.

Kendall, L. (2002). *Hanging out in the virtual pub: Masculinities and relationships online.* Berkeley: University of California Press.

Kouper, I. (2010). Science blogs and public engagement with science: Practices, challenges, and opportunities. *Journal of Science Communication, 9*(1), 1–10.

Lewis, K., Kaufman, J., Gonzalez, M., Wimmer, A., & Christakis, N. (2008). Tastes, ties, and time: A new social network dataset using Facebook.com. *Social Networks, 30*(4), 330–342.

Lynch, M. (2010). Healthy habits or damaging diets: An exploratory study of a food blogging community. *Ecology of Food and Nutrition, 49*(4), 316–335.

Madden, M., & Smith, A. (2010). Reputation management and social media. Pew Internet & American Life Project. http://pewinternet.org/Reports/2010/Reputation-Management.aspx

McKee, H. A., & Porter, J. E. (2009). *The ethics of Internet research: A rhetorical, case-based process.* New York, NY: Peter Lang.

Merz, J. (2010, May 9). Academe hath no fury like a fellow professor deceived. *The Chronicle of Higher Education.* Retrieved from http://chronicle.com/article/Academe-Hath-No-Fury-Like-a/65466/

Morrison, A. (2011). "Suffused by feeling and affect": The intimate public of personal mommy blogging. *Biography, 34*(1), 37–55.

Orgad, S. (2005). *Storytelling online: Talking breast cancer on the internet.* New York, NY: Peter Lang.

Papacharissi, Z., & Gibson, P. L. (2011). Fifteen minutes of privacy: Privacy, sociality, and publicity on social network sites. In S. Trepte & L. Reinecke (Eds.), *Privacy online: Perspectives on privacy and self-disclosure in the social web* (pp. 75–90). New York, NY: Springer-Verlag.

Perry, M. (2011, July 10). Harvard researchers accused of breaching students' privacy: Social-network project shows promise and peril of doing social science online. *The Chronicle of Higher Education.* Retrieved from http://chronicle.com/article/Harvards-Privacy-Meltdown/128166/

Ploderer, B., Howard, S., & Thomas, P. (2010). Collaboration on social network sites: Amateurs, professionals and celebrities. *Computer Supported Cooperative Work (CSCW), 19*(5), 419–455.

Subrahmanyam, K., Garcia, E. C. M., Harsono, L. S., Li, J. S., & Lipana, L. (2009). In their words: Connecting on-line weblogs to developmental processes. *British Journal of Developmental Psychology, 27*(1), 219–245.

Sveningsson Elm, M. (2009). How do various notions of privacy influence decisions in qualitative Internet research? In A. N. Markham & N. K. Baym (Eds.), *Internet inquiry: Conversations about method* (pp. 69–87). Thousand Oaks, CA: Sage.

Tilley, L., & Woodthorpe, K. (2011). Is it the end for anonymity as we know it? A critical examination of the ethical principle of anonymity in the context of 21st century demands on the qualitative researcher. *Qualitative Research, 11*(2), 197–212.

Walker Rettberg, J. (2008). *Blogging.* Cambridge, UK: Polity Press.

Zimmer, M. (2010). "But the data is already public": On the ethics of research in Facebook. *Ethics and Information Technology, 12*(4), 313–325.

Zivkovic, B. (2010). Science blogs and public engagement with science. Retrieved from http://scienceblogs.com/clock/2010/03/science_blogs_and_public_engag.php

CHAPTER THREE

Bridging the Distance

Removing the Technology Buffer and Seeking Consistent Ethical Analysis in Computer Security Research

Katherine J. Carpenter & David Dittrich

Introduction

Computer science (CS) research takes place throughout the United States and the world. When average people think about this type of research, they rarely consider the range of lives touched by access to technology or the ethical concerns raised and still being explored. In the United States, IRBs review most research projects, especially if they are federally funded. Other countries have similar entities called Research Ethics Boards (REBs). Many CS researchers do not perceive their research activities as having human subjects. Academic, military, government, and private institutions sponsor CS research. We focus our discussion on the academic setting, but our points have implications for the military, government, and in the private arena.

Most social or medical research on humans is characterized by the degree of interaction between researchers and research participants, also known as "human subjects of research." Research oversight usually entails a discussion of the harms that could come to human subjects from the research and a plan for monitoring harms once the research begins. In computer security (CompSec) research, the potential participants are difficult to anticipate, and harm to computer end users is not immediately apparent, even though the potential for harm is often fairly high.

In order to combat computer crime, researchers often mimic viruses, botnets, or social engineering attacks. As simulations become more realistic, the possibility of harm to data and computer systems increases. Researcher simulations can compromise computer systems, and data that computer users rely on can be damaged or unavailable, making it unusable.

CS research in general, and CompSec research in particular, often involves manipulating complex systems that have the potential to affect large numbers of individuals. Due to the complicated nature of these interconnected systems, those at risk of harm include computer owners, service providers, and other intermediaries that deliver technological services (Dittrich & Kenneally, 2012). This results in an intricate mixture of potential participants creating confusion regarding the status of CompSec research. Does it qualify as "human subjects research" and require oversight by review boards?

We believe that this type of research needs review, but a shift in research analysis and oversight in this arena is required. Review boards should transition from an informed consent-driven review to a risk-analysis review that addresses potential harms stemming from research in which a researcher does not directly interact with the at-risk individuals. CS researchers may be reluctant to bring forward research that could negatively impact individuals indirectly. CS research is important and should be encouraged. Regulatory reform is needed to provide guidance to researchers and review boards.

Does Computer Science Research Need Review?

Before deciding how to review research, we must first establish that CS researchers undertake research endeavors that actually require review. For example, CompSec researchers may look for vulnerabilities in software so they can repair them. They may mimic viruses and botnets[1] in order to understand them, eliminate them, and identify ways to clean infected machines. Research may include collecting personal data from compromised machines.

"Research means a systematic investigation, including research development, testing and evaluation, designed to develop or contribute to generalizable knowledge" (Protection of Human Subjects, 2005). CS researchers often experiment with goals to improve the technology environment and to break new ground in a rapidly advancing field of technology.

In the United States, when researchers obtain "(1) data through intervention or interaction with the individual, or (2) identifiable private information," they are conducting human subjects research (Protection of Human Subjects, 2005). The federal regulation clarifies as follows: "[i]ntervention includes both physical procedures [...]

and manipulations of the subject or the subject's environment that are performed for research purposes" (Protection of Human Subjects, 2005). Some commenters have suggested there are definitional problems with key terms in federal regulations (Gunsalus et al., 2006, p. 9). We believe the terms "intervention" and "subject's environment" may equally deserve reconsideration and redefinition when it comes to determining greater than minimal risk of harm to humans within CS research.

There is a great deal of CompSec research taking place at universities across the United States, and the majority of it is not reviewed by IRBs that oversee research with human participants. Even when research is reviewed, the review may not be effective if neither the researchers nor the reviewers have the expertise to consider the indirect impact on the end users of technology as "participation" in research (Buchanan, Aycock, Dexter, Dittrich, & Hvizdak, 2011, p. 72). Researchers may not consider the potential harm that the technology could cause to end users, and many review board members are unaware of the potential risks technology users face.

One pillar of ethical research is the informed consent of participants (National Commission for the Protection of Human Subjects of Biomedical and Behavioral Research, 1978). In CompSec research, the range of participants could be anyone with an Internet connection. With such a broad base of potentially impacted individuals, informed consent may not be possible.

The Belmont Report[2] (National Commission for the Protection of Human Subjects of Biomedical and Behavioral Research, 1978) accompanies federal regulations as a guideline for ethical research. The report identifies three ethical principles worthy of consideration in "all human subjects research: respect for persons, beneficence, and justice."[3] We only deal with the first of these, respect for persons, because it encompasses informed consent. Respect for persons is intended to ensure that research participation is voluntary and vulnerable populations are protected. In CompSec research, unless researchers are collecting data, they do not ask for consent because they do not directly interact with anyone (Dittrich & Kenneally, 2011).

We bring attention to and encourage analysis of the harms that can result from this type of research. We recognize that CompSec research is important, and we urge researchers and review boards to consider the resultant harms that could occur to anyone connected to the Internet through their research.

Many understand harm as physical or mental damage or injury (*Merriam-Webster*, n.d.). In the context of CompSec research, it is difficult to cause physical harm to software or computer hardware purely over the Internet. Some potential harm mentioned at the example in the beginning of this section includes compromising private, identifiable information. Identifiable information can be names, identification numbers (including social security or medical record numbers), birthdates, phone numbers, and photographs. The Health Insurance Portability and Accountability Act

(HIPAA) has named 18 identifiers as protected health information, including the aforementioned identifiers, universal resource locators (URLs), and Internet Protocol (IP) addresses. HIPAA is the standard because health research is the most heavily regulated type of research in the United States.

Other possible harms stem from the temporary disruption of computers or networks. If a computer is unresponsive or the network is not available when it is expected to be running, "harm" can be severe—negatively affecting reputation, loss of revenue, disruption of government services or processes, like an election. The more humans rely on information technology as a foundation of their lives, the potential for human harm from research, whether direct or indirect, increases.

Why Are These Protections Insufficient?

If technology is the subject of research, humans who use that technology are not considered "research subjects," but they may be subject to harm directly caused by the research because they use technology. This is true whether the technology is a computer virus or an embedded medical device. Both pieces of technology, viewed in a vacuum, appear to have no impact on humans. In reality, humans interact with both types of technology and are endangered if that technology is disrupted, either on purpose or by accident. Current protections in place for research are insufficient because many IRB members and researchers do not consider end results in research study design. Beyond exposing subject identities, federal research guidelines do not clearly account for research that harms individuals without directly interacting with them.

IRBs Believe They Deal with Distance/Indirection but They Really Don't

Researchers, administrators, and IRB members focus on efficient mechanisms for review. A report from a 2003 Illinois conference recommended, "focusing on those areas of research that pose the greatest risk, such as biomedical research, while removing or reducing scrutiny of many fields within the social sciences and humanities that pose minimal risk" (Gunsalus et al., 2006). This recommendation does not consider whether researchers or IRBs can adequately assess the level of risk that research utilizing Information and Communication Technology (ICT) poses to humans on the Internet, whether those humans are the subjects of research or are simply reliant upon studied ICT resources. We do not advocate increasing the scope of IRB review simply to cover new fields but to accurately identify and review research protocols that pose greater than minimal risk to humans.

Most CompSec research is not reviewed as human subjects research, and if it is, it usually receives review as "minimal risk" or via expedited review procedures that do not require full committee evaluation (Protection of Human Subjects, 2005). Review boards often do not understand the implications of the research or the potential for human harm. IRBs have ample access to medical experts, both on boards and as voluntary consultants, but they often do not have the technical expertise available to evaluate CS or CompSec research protocols. Other commenters have pointed out potential mechanisms for altering the balance of biomedical versus ICT technical expertise by changing the incentive structure for CS researchers so they either sit on IRB committees or at least are known sources of technical expertise for reviewing CS research (Buchanan et al., 2011). This results in a gap of understanding of the actual risks presented by the research protocols. The research review process in the United States focuses on the potential harm to *human subjects* of research, and the research subjects in most CS research studies are technological tools (like computers).

The distance between researchers and harmed humans mentioned earlier perpetuates the lack of research review. Our goal is to bridge the distance between researchers and (potentially) harmed human users of technology by analyzing the potential for harm in varying technological research and building protections into the research design.

Understanding Potential Harms in Computer Security Research

Computer Science and Computer Security Research

CS is a branch of engineering that deals with the theory of computation, the design of computer hardware and software, and the use of computing technology in a variety of fields. It is becoming difficult to find any business, government, personal, or research activity that does not utilize computer systems. We focus on the potential harm that results from a connection to the Internet. Information collection, use, and disclosure (i.e., confidentiality concerns) do not sufficiently address the risks posed by the presence of ICT in research. Risks include breaching confidentiality and compromising the availability and integrity of information and information systems. CS research can be passive—observational—or active. The risk of potential harm is greater in active research. We describe examples of both types here.

Researchers in the subdiscipline of CompSec research attempt to measure and control the tactics, strategies, design, operation, and application of software or computing systems to protect them from malicious or criminal manipulation. CompSec researchers are interested in discovering the vulnerabilities in certain types of software and other technology or in learning how attacks take place and how to prevent them. Malicious software, known as "malware," is the subject of research aimed at reducing the amount of criminal activity that occurs on the internet. Malware can take many forms, but the one we discuss here involves mass infection of third-party computers by criminals to create malicious botnets.

With subcomponents of software engineering, network engineering, and systems engineering, CS is implicitly an engineering discipline. Engineering could be considered a form of experimentation on other humans. Martin and Schinzinger (2005) suggested the following:

> ...engineering should be viewed as an experimental process. It is not, of course, an experiment conducted solely in a laboratory under controlled conditions. Rather, it is an experiment on a social scale involving human subjects.

Engineering may be considered experimentation, but most engineering falls outside the purview of federal research regulations, including CS research. Academic researchers who test technology fail to recognize the way their research constitutes experimenting on humans.

Examples of Computer Security Research

Many types of research fall under the umbrella of "computer security" research. We address four ethically challenging forms.

Vulnerability research. Vulnerability research identifies technical weaknesses in systems that could allow an adversary to bypass security mechanisms and assume control or elevate privileges within a protected system. This research is important because software vulnerabilities are exploited to gain control of computers through an infection mechanism that creates bots and botnets or to find avenues that could be used to disrupt or disable building control systems, embedded medical devices, automobile control systems, or electronic voting systems.

There is a long-standing controversy surrounding vulnerability disclosure, with two extreme viewpoints: full disclosure of vulnerabilities upon discovery and nondisclosure. Researchers may publicly disclose information before contacting the producers of vulnerable systems. This early disclosure is based on the researcher's desire to get maximum press exposure, to avoid legal pressure from the affected vendor. Researchers may believe affected vendors will not act unless the public is fully informed about the ways the vulnerability could be exploited to cause harm, and disclosure engages market forces that compel the vendor to respond.

Formalized vulnerability disclosure guidelines go back to the early 2000s. Carnegie Mellon University's CERT/CC[4] was one of the first programs to publish vulnerability disclosure guidelines. Other guidelines came from the Organization for Internet Safety, the National Infrastructure Advisory Council, and CompSec firms and software vendors over the years (Dittrich & Kenneally, 2012).

A high profile example of a successful coordinated disclosure of fundamental vulnerabilities in the Domain Name System (DNS) services relied upon by all Internet users occurred in 2008 (Kaminsky, 2008). Hundreds of vendors were involved over several months to ensure that the majority of them were prepared to release patches fixing the problem on or about the same time as a coordinated public disclosure occurred. A blog post by Moussouris and Stone (2009) publicly acknowledged Microsoft's coordinated vulnerability disclosure activities and processes. A full statement of the policy indicating how Microsoft could assist security researchers with coordinated disclosure was released in 2011 (Microsoft Security Response Center, 2011).

Botnet takedown research. Researchers at universities, security software vendors globally, and private individuals actively engage in botnet research and takedown activities. These actions benefit society by taking stolen computing assets out of the hands of miscreants who may be causing significant financial harm. Effectively, botnet takedowns become "good guys" fighting with "bad guys" to control stolen computing assets belonging to innocent victims who may be unaware their computers are compromised. These takedowns pose additional risks to computers infected with malware above and beyond monitoring risks in passive network traffic monitoring studies.

The Department of Commerce and National Institute of Standards and Technology (NIST) published a commerce notice in the Federal Register promoting "Models to Advance Voluntary Corporate Notification to Consumers Regarding the Illicit Use of Computer Equipment by Botnets and Related Malware" (Department of Commerce, 2011). This notice cites researchers who "suggest an average of about 4 million new botnet infections occur every month." They mention other efforts to involve Internet Service Providers (ISPs) in helping clean up infected computers, such as the Internet Engineering Task Force (IETF) draft, "Recommendation for the Remediation of Bots in ISP Networks" (Livingood, Mody, & O'Reirdan, 2010). The notice indicates several international governmental programs dedicated to preventing botnet infections from spreading. Botnet infections are a public threat and the public is served by preventing them.

The commerce notice pointed out that, "Internet Service Providers (ISPs) [have] contact information for the end-user and a pre-existing relationship" with them that could facilitate communicating with end users. Utilizing the ISP relationship may not be a suitable mechanism for obtaining informed consent, as the number of users involved is still large enough as to make it impractical. It would be a costly mechanism to obtain informed consent from all end users. However, we recognize that ISPs *could* act as proxies of consent for their clients (Dittrich & Kenneally, 2012).

The public wants to feel safe. Public pressure may encourage researchers to clean up infected computers without involving end users. Recently, Microsoft obtained an ex parte temporary restraining order allowing the company to disable thousands of malicious domain names and sinkhole the Kelihos botnet with Kaspersky Labs researchers (Boscovich, 2011). An unscientific poll by Kaspersky[5] asked the following: "How should Kaspersky proceed with the Hlux/Kelihos Botnet?" Eighty-three percent of respondents believed Kaspersky should "push a cleanup tool that removes the infections" as opposed to doing nothing or working with ISPs to notify infected victims. Pressure from a frustrated general public, combined with a lack of generally accepted guidelines for researchers to consider such risky actions, may push researchers to take high risks that could potentially damage computers or data in their effort to stem what the public views as widespread harm by criminals.

Electronic voting research. The electronic vote tabulation systems used in many states employ computer hardware that is very similar to the personal computers in our homes, running similar, popular, commercial operating systems on custom software that presents options to a voter. These computers record accumulated votes and deliver them to a central tabulation system that provides the election results using network connections. Several research groups have examined both the hardware and software of these systems to prevent elections occurring on flawed equipment.

Elections can be, and have been, manipulated by those who wish to serve their political interests. Flaws that can be remotely exploited like manipulating votes, altering counts, and bypassing audit checks pose a serious threat to the integrity of the voting system.

Cyber-physical systems research. The term "cyber-physical system" (CPS) broadly applies to any type of computing device that can be controlled remotely through a computer control system and has direct physical effects on objects in the physical world. This can include process control systems in factories, embedded medical devices that control heart rhythm or delivery of drugs like insulin, and brake systems in automobiles. Research on these types of devices would almost never undergo IRB review because humans are not involved when researchers simply study the device's function outside of their intended use. A vulnerability discovered in a CPS that can be exploited remotely, using low-cost equipment, could allow a malicious actor to control these systems and literally kill someone.

Refining Our Idea of Distance

The end users of tested technology are people, and if or when harm occurs to the computer or system because of the research, the human user is often affected negatively. This distance between researcher and affected individual indicates that a paradigm shift is necessary in the research arena. We support the notion that an ideological transition in research regulation from "human subjects research" to "human harming research" is necessary (Aycock, Buchanan, Dexter, & Dittrich, 2012; Dittrich & Kenneally, 2011, 2012).

When ICT is present in a research setting, the risk of harm may be much broader than in direct intervention research, where the research subject is the primary party at risk. This is illustrated in Figure 3.1.

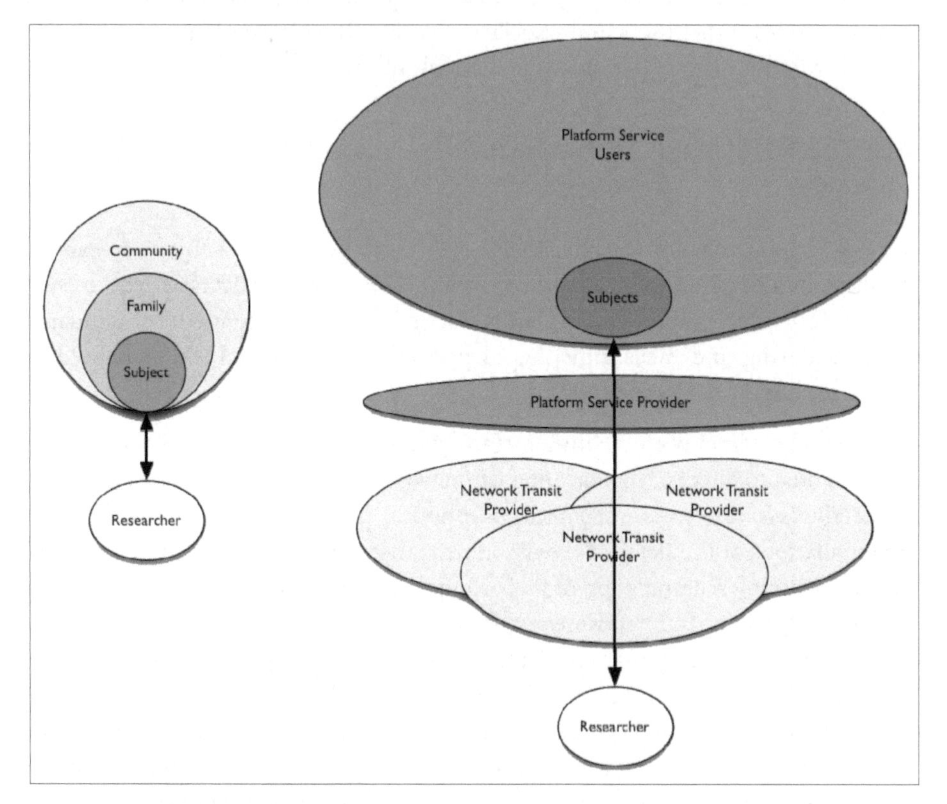

Figure 3.1. Relationships among research, subject, and other stakeholders.

The arrow in Figure 3.1 illustrates the relationships between researcher and "human subject."The amount of gray indicates relative level of potential risk (lighter meaning less risk, darker meaning more risk) to stakeholders.

The left side of Figure 3.1 shows the risk relationships in a typical biomedical or behavioral research study. In this example, the researcher interacts directly with the subject and potential harm extends indirectly to other parties and decreases in severity (although not necessarily in total number of those impacted) as you move away from the subject through that individual's relationships.

The right side of Figure 3.1 shows the risk relationships in a study that uses ICT. In researcher-subject interaction, intermediary providers of ICT services and the subjects are a subset of the entire user base of a given platform provider (e.g., a social network site). Notice how the degree of risk is inverted from the previous example. There may be little or no risk to transit providers, but the platform provider and/or other users in the social network of the subjects being studied may be at nearly the same risk level as the research subjects.

Many CS Researchers Do Not Believe Their Research Is Human Subjects Research

CS researchers may not consider their activities to be "human subjects research," resulting in a dearth of regular and consistent assessment of adverse effects to systems or individuals. The CS community does not have consistent ethical standards for considering and measuring the adverse effects of research on stakeholders (Dittrich, Bailey, & Dietrich, 2009).

People interact with technology (and vice-versa). CS and CompSec research is not limited to direct or manual interaction with human end users; it engages them indirectly. *Intervention* is not limited to physical procedures. Intervention can be "manipulations of the [subject's] environment that are performed for research purposes," including manipulation of their computing devices or automated appliances at home (Office for Human Research Protections, 2008).

Interactions with humans in virtual environments using ICT impact exponentially more individuals than a direct intervention would. Consider a behavioral study performed within an online virtual world environment observing avatar interactions. Some may argue an avatar is a graphical construct representing a human, but it is not a "living individual" who can "interact" with the researcher and thus is never subject to ethical review. The same argument is used for researching malicious software, embedded medical devices, or a process control device in a dam preventing water from flooding cities. Despite the fact that humans are not the *direct subjects* of the research, the research may involve greater than minimal risk to humans.

CS and CompSec research does not seem like human subjects research because technology appears to act as a buffer between researchers and individuals. This buffer seems present even though research may directly impact individual users of a system or compromise their computer. Research is generally designed to be as transparent as possible at least for the researchers and their funding agencies. The nature of ICT research diminishes transparency and creates a distance between the researchers and potentially impacted parties (be they direct participants or indirectly involved "research subjects," because ICT itself is the subject of research; Buchanan et al., 2011).

The following diagram (see Figure 3.2) compares three research contexts to illustrate this concept.

In the Figure 3.2, the term "research object" may be a more applicable term than "research subject," if a researcher is trying to separate the focus of a research activity and the individual who may be harmed.

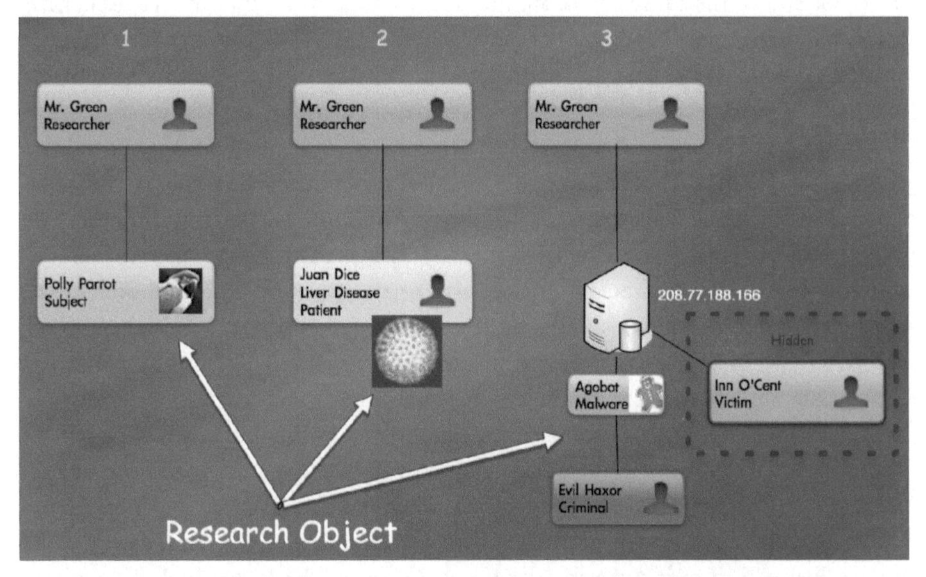

Figure 3.2. Relationship between research and subject/object of research.

Passive research. A researcher directly observes his subject in Context 1. These observations are passive and do not require direct intervention. If the observed behavior is noncontroversial, in a public space, and no personally identifiable information is obtained, the risk to subjects is minimal or nonexistent. Participants may not even need to provide consent.

Active intervention. In Context 2, a researcher studies disease organisms in the liver of a living human (i.e., the "research subject"). The organism is the object of the study. In order to observe the liver, the researcher must perform experimental procedure(s) on the human owner of the liver. This entails both informed consent (permission) and possibly some risk of harm to the human "subject." The researcher must directly interact with the subject. Obtaining informed consent is facilitated by conducting face-to-face interaction with the researcher, engaging in dialogue, asking questions, and actively participating in research procedures. It is clear who is at risk, what those risks are, and whether those risks materialize as harm.

The final situation, Context 3, indicates the studied object is either a piece of malicious software (e.g., a "bot") or a criminal who infected a third party's computer and is engaged in illegal activity. A researcher may directly intervene with the malicious bot software, but the owner and other users of the infected computer are completely unaware of the bot software, the infection, and either the criminal or researcher's manipulation of their infected computer. The best identifier that the researcher has is an IP address. Some regulations consider an IP address to be an identifier. However, an IP address does not allow the researcher to identify or communicate with the owner and users of the infected computer any more than having a street address guarantees that the researcher can find and communicate with a specific individual at a residence (Ohm, 2009). This makes obtaining informed consent impractical at best, or impossible at worst, when the research may discover millions of infected computers' IP addresses. The risk that must be balanced is not risk to the object of the research (i.e., the bot) but to the humans who own and/or use the infected computer. If the researcher's manipulations cause the accidental destruction of data within the infected computer, the harm is indeterminate and possibly unquantifiable (yet to the person whose data are affected, significant and irreparable harm may occur).

If research involves passive observation, the risk to the humans may only consist of harms that occur from disclosure of their personal identity, behavior, or participation as a research subject. The risk of harm is revealing confidential information about the human. When the research activity is directly active, the harm can be physical or financial, even if a private user is never exposed. Risks can include compromising the availability and integrity of information, along with the systems that contain information. The potential harms can expand beyond the primary owner and/or users of the computer system. Secondary harms are likely if individuals rely on the same information bases and/or information systems.

Situations 2 and 3 differ dramatically in terms of informed consent. In Situation 2, the human has given consent, accepted the risk of participating in research, and is aware of possible causes of harm. In Situation 3, harmed individuals cannot know the cause. They are unaware that their computer is involved in research activity. Harm simply occurs and the harmed humans must guess the cause. It is easier for an individual to blame a criminal or buggy software when they suffer a computer malfunction. If a researcher causes harm, the researchers and their institution may suffer public ridicule or reputational harm.

Spatial distance. The digital world can affect anyone connected to it no matter where they are in the physical world. The digital arena allows us to do wonderful things, but the harm inflicted through digital technology on companies that do not protect themselves from attack or unsuspecting individuals can be tremendous. We are all familiar with the concept of "distance" between two objects in physical space. Distance also manifests in a logical sense and a temporal sense.

Logical distance. The concept of logical distance[6] refers to the hierarchical relationships within ICT systems. These relationships form a graph with multiple intermediate levels. End users can be viewed as nodes at the edges of this graph. They obtain ICT services from platform or application providers, who rely on foundational computing resource service providers and network transport providers. All of these providers are interdependent, and the failure of any one can cause failures in others, with the end result that an application end users depend on does not work. The more layers there are in this provider/consumer hierarchy, the greater the logical distance between the researcher and the end users who may be affected by disruption of a service being subjected to an experiment by a researcher. In other words, an action taken by a researcher that affects a provider may affect a company that consumes the provider's services and sells other services to secondary providers, who have their own customers who may be affected by disruption to the upstream provider in a cascading manner.

Temporal distance. Humans tend to consider actions and their effects in immediate terms, but the beneficial effects and harms of research or knowledge do not always occur simultaneously. For this reason, human calculations of risk may be inaccurate. Software producers can use vulnerability knowledge to fix the problem and enhance reliability in the software system. The same knowledge can cause harm. Developing and testing patches, notifying customers or computer users, and downloading and implementing the patch takes time. In the gap, malicious actors can take advantage of those vulnerabilities to penetrate an unprotected system and either steal personal data or compromise the computer to distribute viruses.

Vulnerabilities affecting over 150 vendors were discovered in the DNS in 2008 (Faber, 2008). Public disclosure of the vulnerability occurred 6 months after the initial discovery. During those 6 months, vendors were notified and attempts were made to maximize the number of vendors who could provide a mitigating patch for the problem upon public notification. When the vulnerability was revealed, providers and consumers raced to patch their systems before malicious actors could exploit the vulnerabilities harmfully. Calculating vulnerability risk requires a harm function that sums harm over time and benefit over time (Arora & Telang, 2005). When a researcher publishes vulnerability information, the resulting harm may continue long after publication occurs, not just immediately afterward.

Where Do We Go from Here?

Next Steps

We need to solve two fundamental problems to improve the situation we have just described. First, we must facilitate the identification of humans who could be impacted in some way by using ICT that is the subject of research or is used by CS researchers as a conduit for interaction or intervention with humans. Without this comprehensive understanding of human involvement, potential causes of harm will be underestimated or be left out of the equation altogether. Second, we need to shift the focus of ethical review onto those situations in which there is a greater than minimal risk to humans from research activities themselves, not on whether there is direct intervention with humans as the subjects of research.

CS researchers want to do good for society, but good intention alone may not be enough. Ethicist Annett Markham (2006, p. 38) claimed that ethics can be the basis for conscious decision making about research methodology that reflects one's intentions and his or her source of "consciousness, mindfulness, honesty, and sensitivity," suggesting self-reflective questions for researchers: "What is the purpose of this research project?" "What is the potential or desired outcome and why is research necessary to this outcome?" (Dittrich, Leder, & Werner, 2010, p. 2). It may be important to ask how stakeholders interpret researcher actions. Will stakeholders feel grateful, neutral, or resentful of researchers?

Balancing Risks and Benefits of Research Using Stakeholder Analysis

In vulnerability research, it is easy to identify with system users and take their side against the creators of these systems whose design flaws may put users at risk. The justification for releasing vulnerability information and proof-of-concept (POC) exploit software/hardware often focuses only on a general claim that "the public deserves to know about vulnerabilities in order to fix them," or to motivate vendors to fix vulnerable products.

Rarely do such justifications go beyond consideration of the general category of "the public" and include the full range of specific stakeholder groups (both those to be protected and those intent on using such information to do harm.) In part, this is because quantifying costs and benefits for just these two stakeholders can be difficult. It is also not part of the research itself, which is focused on the system and its vulnerabilities. Yet in reality there are even more stakeholders who may be affected by publication of vulnerability research, which must be considered as potential harms related to research activities and dealt with in evaluating research ethics along with any other harms and benefits. There is no widely recognized and adopted standard methodology in the CompSec community; however, there are examples of this technique being used (Dittrich & Kenneally, 2011, 2012; Dittrich et al., 2010).

Stakeholders include researchers, human subjects, society, and criminals/attackers. These broad categories can become researchers and their programmers; vendors who use the Internet to sell products or ISPs; the programmers for vendors; clients and customers of websites, online stores, and ISPs; and criminals who exploit Internet-based services and/or the data that they are able to discover through technological vulnerabilities. Some academic research activities can look like criminal activity, even though they exist to track an individual's reactions to attacks and not to exploit personal data (e.g., stealing credit card information).

In several research cases involving electronic voting machines, researchers have chosen to publicly disclose vulnerability information before notifying anyone. They feared a massive legal backlash from voting machine vendors. They believed it was necessary to inform the public so they could continue to trust the integrity of the election process. Tremendous political and financial stakes are at risk with the potential of voting fraud. Vendors are motivated to refute vulnerabilities, and researchers can gain positive publicity by exposing technological flaws. The public is the primary beneficiary of disclosure and mitigation of voting systems' vulnerabilities. Many other stakeholders are involved in the certification, purchase, operation, and monitoring of voting machines and election results. There is little or no discussion of the positive or negative effects on other impacted stakeholders, such as the following:

- **Local municipal governments:** These government entities store, transport, and operate voting systems. They function on fixed annual budgets. They may be elected (accountable directly to the voters) or appointed (accountable to elected officials). They rely on volunteers who help operate polling locations.
- **State and federal legislatures:** Legislatures appropriate funds to purchase or maintain voting systems. Replacing voting machines is costly and it is time-consuming to appropriate funds that make these systems available or replace them. Help America Vote Act (HAVA) money was

spent years ago, and funding to replace voting systems it purchased for the states must come from future federal appropriations (unlikely in the current political climate) or state funds (which are similarly politically difficult to obtain presently).

- **State government executives** (i.e., secretaries of state): State executives establish certification standards for voting systems. All systems must pass an evaluation based on existing (or newly defined) standards. Evaluations take months or years to complete.
- **Citizens:** Citizens want clean and fair elections and trust in the voting process. Citizen watchdog groups monitor government activities and hold their representatives accountable for action or inaction. Private citizens volunteer as election monitors in many precincts to ensure that votes are properly handled and tabulated. These groups may assist in reforming the voting system or may actively oppose elected officials who they believe do not ensure a trusted voting system.
- **Law enforcement agencies:** Law enforcement agencies at the state and federal level enforce voting laws. If a voting system is questioned so near an election that reverting to paper ballots is impossible, tensions about election fraud may arise. Doubt about election validity can incur significant costs in both civil and criminal legal actions. In *Bush v. Gore* in 2000, the civil legal process moved forward at a "deliberate pace," which did not fit mandated time frames for recounts or challenges to election results.

It is challenging to identify a means of notification that balances the benefits and risks to all stakeholders. There may be risks beyond the control of researchers and it may be difficult to identify an entity who will respond to reports of serious flaws in the voting system and who can assist in remediation or implementation of additional audit mechanisms during an election with flawed equipment.

The timing of vulnerability announcements plays a large role in potential harm that could take place if an election occurred using flawed equipment. Factors are the speed with which individual stakeholders can act, the fixed time frames of elections in the United States, and the balance of vulnerability information disclosed against the mitigation information that could secure the integrity of the voting system. Over time, trusted venues may exist in which researchers can publish sensitive information. Researchers may become more familiar with coordinated vulnerability disclosure mechanisms that are well positioned to coordinate sensitive disclosure and mitigation efforts.

Shift to Regulating Research That Causes Harm to Humans Rather Than Focusing on the Harm Specifically to Human Subjects of Research

We support the idea of a paradigm shift from assessing "human subjects research" to assessing "human harming research" in the CompSec research arena (Aycock et al., 2012). The idea that the only source of research harm stems from direct interactions between individuals is obsolete in research fields that connect individuals and machines globally. Similarly, focusing only on data and identifiability of individuals who are the subjects of research leaves out other types of nontrivial harms.

The distance between technical expertise available to IRBs and how researchers portray risks in their research protocols is beyond the scope of this chapter, but it has been noted by others (Gunsalus et al., 2006; Borenstein, 2008). Borenstein explained that, "[t]he goal of protecting human subjects should not be brushed aside merely so that researchers can proceed with their work . . . [A]n effective review can detect flaws and prevent limited resources from being wasted" (Borenstein, 2008, p. 5). The goal of an effective review of the technological issues cannot be achieved without available expertise on IRBs.

Recommendations to improve the efficiency of IRB review suggest allowing researchers to use established standards to bypass IRB review altogether (Department of Health and Human Services, 2011). If researchers remove all data in the list of 18 HIPAA identifiers, current research can be exempt from review. Anonymizing data increases privacy and encourages expedited review of research. However the techniques used to anonymize data are generally not sufficient to prevent re-identification of data (Ohm, 2009, p. 1701). The standards-based checklist only addresses privacy risks associated with data and ignores risks to integrity or availability of information or information systems.

ICT is constantly evolving, converging, and becoming more ubiquitous and transparent. As this occurs, ICT starts to be taken for granted the same way that those who live in the developed world rarely give a second thought to roads, electricity, or drinking water. The result is greater and greater chances that disruption or damage of ICT components will have human-harming effects.

We would like to begin an ideological shift toward regulating research that could cause harm to humans as an alternative to the current regulatory bent that focuses solely on the harm to *human subjects of research.* In the CS world, it is difficult to determine who is *actually* affected by CS research, and it is better to consider the potential for harm when designing research and create a strategy to minimize the potential for harm and mitigate harm when it occurs. This also achieves the goal of optimizing ethical review of all research that involves ICT, including data-intensive studies in the biomedical research field, as the same type of analysis of stakeholders and potential benefits and harms to them are identical. We believe that over time, the capacity of IRBs to efficiently perform their valuable oversight duties, and ability of CS researchers to efficiently structure their research protocols and present them to IRBs for review, will bridge the distance and work toward consistent ethical analysis in CS Research.

Acknowledgments

The authors would like to thank Michael Bailey and Erin Kenneally for generously allowing their conversations and correspondence to help in writing this work, to participants in the Menlo Report working group for spirited debates about ethics in ICT research, to Elizabeth Buchanan and Laura Odwazny for exploring the concept of "distance" as it applies to Internet research under the Common Rule, and to Halle Showalter Salas for encouraging simplicity.

Notes

1. Individual infected computers (known as "bots," short for "robot") form distributed malware systems used by criminals (known collectively as a "botnet," short for "robot network").
2. The National Commission for the Protection of Human Subjects of Biomedical and Behavioral Research published "Ethical Principles and Guidelines for the Protection of Human Subjects of Research" in 1978. This report is commonly called "The Belmont Report" after the Belmont Conference Center where the commission met to discuss and draft the report.
3. Quoted from the description of the Belmont Report in "Human Subjects Research (45 CFR 46)" on the Department of Health and Human Services' web site: http://www.hhs.gov/ohrp/humansubjects/guidance/
4. Originally "Computer Emergency Response Team," the name changed to "CERT/CC [TM]" because the coordinating center (CC) was the brand they preferred to maintain (http://www.cert.org).
5. A total of 4,329 individual responses were recorded in this unscientific, blog-based poll as of January 1, 2012 (see http://www.securelist.com/en/polls?viewpoll=207796946).
6. Markham and Buchanan use the term "conceptual or experiential distance" to discuss this concept (Buchanan et al., 2011).

References

Arora, A., & Telang, R. (2005). Economics of software vulnerability disclosure. *IEEE Security & Privacy, 3*(1), 20–25.

Aycock, J., Buchanan, E., Dexter, S., & Dittrich, D. (2012). Human subjects, agents, or bots: Current issues in ethics and computer security research. In G. Danezis, S. Dietrich, & K. Sako

(Eds.), *Financial cryptography and data security* (Vol. LNCS 7126 of Lecture Notes in Computer Science, pp. 138–145). Berlin, Germany: Springer-Verlag.

Borenstein, J. (2008). The expanding purview: Institutional review boards and the review of human subjects research. *Journal of Clinical Research Best Practices, 5*(2), 5. Retrieved from http://first-clinical.com/journal/2009/0902_AIR_Purview.pdf

Boscovich, R. (2011). Microsoft neutralizes Kelihos botnet, names defendant in case. Retrieved from http://blogs.technet.com/b/microsoft_blog/archive/2011/09/27/microsoft-neutralizes-kelihos-botnet-names-defendant-in-case.aspx

Buchanan, E., Aycock, J., Dexter, S., Dittrich, D., & Hvizdak, E. (2011). Computer science security research and human subjects: Emerging considerations for research ethics boards. *Journal of Empirical Research on Human Research Ethics: An International Journal, 6*(2), 71–83.

Department of Commerce and Department of Homeland Security. Models to Advance Voluntary Corporate Notification to Consumers Regarding the Illicit Use of Computer Equipment by Botnets and Related Malware. 76 Fed. Reg. 58466 (2011).

Department of Health and Human Services (DHHS). (2011). Information related to advanced notice of proposed rulemaking (ANPRM) for revisions to the Common Rule. Retrieved from http://www.hhs.gov/ohrp/humansubjects/anprm2011page.html

Dittrich, D., Bailey, M., & Dietrich, S. (2009). *Towards community standards for ethical behavior in computer security research* (Stevens CS Technical Report 2009–1). Stevens Institutue of Technology, Hoboken, NJ, USA. Retrieved from http://www.cs.stevens.edu/%7Espock/pubs/dbd2009tr1.pdf

Dittrich, D., & Kenneally, E. (Eds.). (2011). *The Menlo Report: Ethical principles guiding information and communication technology research.* Retrieved from http://www.cyber.st.dhs.gov/wp-content/uploads/2011/12/MenloPrinciplesCORE-20110915-r560.pdf

Dittrich, D., & Kenneally, E. (Eds). (2012). *Applying ethical principles to information and communication technology research: A companion to the Department of Homeland Security Menlo Report.* Available at http://www.cyber.st.dhs.gov/

Dittrich, D., Leder, F., & Werner, T. (2010, January). *A case study in ethical decision making regarding remote mitigation of botnets.* Paper presented at the Workshop on Ethics in Computer Security (WECSR '10), Tenerife, Canary Islands, Spain.

Faber, S. (2008, September). *Responsible disclosure: A case study of CERT VU #800133, "DNS cache poisoning issue."* Paper presented at the workshop of the Domain Name System Operations Analysis and Research Center (DNS-OARC), Ottowa, Ontario, Canada.

Gunsalus, C., Bruner, E., Burbules, N., Dash, L., Finkin, M., Goldberg, J., Greenough, W., Miller, G., & Pratt, M. (2006). *Illinois whitepaper—improving the system for protecting human subjects: Counteracting IRB mission creep* (Report No. LE06–016). University of Illinois College of Law, Champaign, IL, USA. Retrieved from http://ssrn.com/abstract=902995

Kaminsky, D. (2008, August). *Black Ops 2008—it's the end of the cache as we know it.* Presented at Black Hat Briefings USA 2008, Las Vegas, Nevada, USA.

Livingood, J., Mody, N., & O'Reirdan, M. (2010). *Draft recommendations for the remediation of bots in ISP networks.* Retrieved from http://tools.ietf.org/html/draft-oreirdan-mody-bot-remediation-03

Markham, A. (2006). Ethic as method, method as ethic: A case for reflexivity in qualitative ICT research. *Journal of Information Ethics, 15*(2): 37–54.

Martin, M. W., & Schinzinger, R. (2005). *Ethics in engineering* (4th ed.). New York, NY: McGraw-Hill.

Merriam-Webster Dictionary and Thesaurus Online. (n.d.) "Harm." Retrieved from http://www.merriam-webster.com/dictionary/harm

Microsoft Security Response Center. (2011). Coordinated vulnerability disclosure at Microsoft. Retrieved from http://go.microsoft.com/?linkid=9770197

Moussouris, K., & Stone, A. (2009). Threat complexity requires new levels of collaboration. Retrieved from http://blogs.technet.com/ecostrat/archive/2009/07/27/threat-complexity-requires-new-levels-of-collaboration.aspx

National Commission for the Protection of Human Subjects of Biomedical and Behavioral Research (1978). Ethical principles and guidelines for the protection of human subjects of research (DHEW Publication No. [OS] 78–0008). Washington, DC: U.S. Government Printing Office.

Office for Human Research Protections (OHRP). (2008). *Guidance on engagement of institutions in human subjects research.* Retrieved from http://www.hhs.gov/ohrp/policy/engage08.html

Ohm, P. (2009). Broken promises of privacy: Responding to the surprising failure of anonymization. *UCLA Law Review, 57,* 1701. Retrieved from http://ssrn.com/abstract=1450006

Protection of Human Subjects, 45 C.F.R. pts. 46 102(d) & 102(f) (2005).

Section 2

Digital Ethics in Practice

Introduction by Don Heider

Why worry about digital ethics? Because new technologies have brought with them new questions. In the sections that follow, authors deal with some of these new questions and new issues that have arisen as a result. Traditional ethical guidelines might offer some help, but simply applying them de facto does not always work well. Ethics at its most basic form is a set of moral principles that help guide our behavior. In my own experience, ethics is more of a process than any finite set of rules. It's in the asking of difficult questions, in the consideration of sticky situations, in taking into account how our behavior might affect others, that some clarity emerges. But for me, ethics is more about wrestling in the gray area rather seeing clear black and white.

In the first piece in this section, David Kamerer takes a look at online publishing and the subtle ways in which writers may endorse products or services without making clear their connections with those products and services. In other words, is it ethical to get paid to write about and endorse a product without letting your audience know about the compensation involved? In recent times this has become more than an ethical question; in the United States it also involves the Federal Trade Commission, which David explores in detail in the piece.

The Internet has certainly given many more people the opportunity to let their feelings known through blogs and other ways of posting. But it has also given everyday citizens a chance to try their hand at journalism. Jessica Roberts and Linda Steiner look at the phenomenon of what's called citizen journalism and sites designed specifically to give people a chance to report and write about events and news. But what ethical and journalistic guidelines do these sites offer to these contributors who may have no background in journalism as a profession? And what new ethical questions arise as a result of citizens trying their hand at journalism? These questions are discussed in what follows.

When it comes to posting on the Internet, it's not just people's words that raise interesting new questions; it's also pictures. Mark Grabowski and Sokthan Yeng are concerned about the rise of a new breed of website which concentrates on posting mug shots—those photos shot by law enforcement officials when a suspect is arrested and charged with a crime. Newspapers have different standards and guidelines for publishing such photos, but as we've come to understand, things posted on the Web have a much longer shelf life, raising more serious questions of whether to post these mug shots, and, given that a number of people who are arrested are eventually acquitted, when the mug shots should be taken off the Web. Professors Grabowski and Yeng discuss the ethical consideration of these mug shot websites.

Photos and words and other media artifacts such as videos, audio clips, and so forth, that are posted by individuals shape a person's image on the Web. Erin Reilly looks at how living our lives in this very public forum of the Web allows us to build an online identity, but it also may have some unintended consequences. When it comes to posting on behalf of children, some difficult questions come up as to what role the children have in this process. This piece looks at the shadows cast by all these postings and how those might have long-term consequences.

Together these pieces ask some difficult and important questions about the new landscape of the Internet and the ways in which this new technology raises important new ethical considerations.

CHAPTER FOUR

Disclosing Material Connections Online

Legal and Ethical Issues

DAVID KAMERER

Aristotle, the father of rhetoric, identified three great appeals: logos, or logic; pathos, or emotion; and ethos, the appeal of reputation (Heinrichs, 2007). Ethos is why professional athletes are paid to sell running shoes and why actresses are paid to sell fragrance and fashion. The communicator can greatly influence the message, adding legitimacy and elevating the product in the mind of the audience.

In an ad, it's clear that the actress is paid to sell the fragrance. Online, things are much murkier. When you read an endorsement of a product on Yelp, Facebook, or on a blog, it can be difficult to tell what relationship the writer has with the product or service in question.

If the endorser is paid to endorse the product, he or she is said to have a material connection to the product or service. Our culture has long held that when people are paid to endorse something, they should disclose it. For example, broadcasters have been fined for accepting money to play particular records ("payola"), failing to identify the sponsor of a video news release, or weaving commercial messages into programs without notifying the audience (Payola and Sponsorship Identification, n.d.)

While broadcasters have an explicit economic model upon which to run their ventures, many people who self-publish online in blogs or social networks do not. People who publish online do so for many different reasons—hobbyist, political enthusiast, or simply self-appointed expert. These publishers are culturally diverse, may not be formally trained as communicators, and may not subscribe to self-regulation or follow a particular code of ethics. As a result, many online publishers accept some form of payment (including free products, free samples, free trips, review copies, or cash payment) without disclosing these payments to their readers.

Complaints from citizens about commercial speech attracted the attention of the Federal Trade Commission (FTC) as far back as the mid-1970s. The FTC most recently formally addressed these concerns in 1980, with its "Guides Concerning the Use of Endorsements and Testimonials in Advertising"(FTC Publishes Final Guides, n.d., a). Those guidelines remained unchanged for 30 years, while the media landscape underwent a digital transformation. In 2009, the rules were amended, in part to rein in the practice of nondisclosure of material connections online and within social networks.

The Federal Trade Commission and Disclosure Rules

The FTC is an administrative agency that was created by Congress in 1914 to promote competition in commerce. The original impetus of the FTC was to "bust the trusts" that became prominent in the late 19th century. The agency has investigative and enforcement functions in the general area of consumer protection. The FTC has authority to regulate disclosure of material connections under its charge to regulate unfair and deceptive advertising practices (Federal Trade Commission—About Us, n.d.).

In the fall of 2009, the FTC updated its Guides Concerning the Use of Endorsements and Testimonials in Advertising (FTC Publishes Final Guides, n.d., a). The guidelines became effective on December 1, 2009. The 81-page document identified three main updates to the FTC's rules:

- Restrictions on describing "typical" results in advertising.
- Guidelines for celebrity product endorsers.
- Guidelines for disclosing material connections between advertisers and endorsers.

Of these three, the "material connections" guidelines are the subject of this research. The release of these guidelines created a stir in the blogosphere and social media space, a largely unregulated marketplace of ideas, commerce, and hucksterism. There was talk of $11,000 fines for bloggers who ran afoul of these regulations, which turned out to not be true (FTC Responds to Blogger Fears, n.d.).

The announcement of the new guides generated much discussion among bloggers, in large part depending on their position in it. Established bloggers, for example, said the ruling would cut down on the "cloggers," wanna-be influencers who were blogging just to get free products and services (New F.T.C. Rule Has Bloggers, n.d.). Some said it was an important step forward that would improve the credibility of the blogosphere.

FTC Actions

While even a "raised eyebrow" from an administrative agency may have a chilling effect on free speech, so far the FTC's actions are modest. Thus far the agency has considered or discussed six events, levying a fine in one instance, settling in two with requirements for ongoing reporting, and taking no action in three others. The incidents involve apparel retailer Ann Taylor Loft, public relations firm Reverb Communications, online marketer Legacy Learning Systems, rapper and entrepreneur 50 Cent, actor and entrepreneur Ashton Kutcher, and Hyundai automobiles.

Ann Taylor Loft. In January 2010, Ann Taylor Loft invited 31 bloggers to see the new summer clothing line. The bloggers were encouraged to write about the new clothing and then submit their content to a public relations person in order to win a gift card. The value of the gift card was not revealed until after the submission. According to http://jezebel.com, the note said the following:

> Come take a sneak peak at LOFT's summer 2010 collection before anyone else! . . . Bloggers who attend will receive a special gift, and those who post coverage from the event will be entered in a mystery gift card drawing where you can win up to $500 at LOFT! (Fashion Bloggers Run Afoul, n.d.)

Coverage from the bloggers was very positive. Only 2 of the 31 attendees disclosed the gift from Ann Taylor Loft. This aspect of the promotion received some blowback from critics. In addition to http://jezebel.com, other blogs and press outlets publicized the controversy, including the *Los Angeles Times* blog (questioning the journalistic ethics of Ann Taylor Loft's practices) and then *PR Week,* questioning whether the FTC would take action based upon what were then recent updates to the endorsements' guidelines, which had only been in effect for a few weeks at the time of the event in question.

The FTC did take notice and slapped Ann Taylor Loft on the wrist, noting that this was a first-time occurrence, that the company had agreed to go over disclosure rules with bloggers in the future, and that the effect of the promotion was small. The agency's conclusion was as follows:

> Upon careful review of this matter, we have determined to not recommend enforcement at this time. (FTC Gives Ann Taylor a Pass, n.d.)

Here is the takeaway: It is primarily the advertiser's responsibility, not the blogger's, to make sure the material connection is disclosed. The advertiser's responsibility is substantial. The advertiser needs to inform the blogger of a disclosure requirement, document process, monitor for compliance, and be able to identify and remove content that is not disclosed and/or influencers who continue to fail to disclose.

Reverb Communications. The second FTC action resulted in a settlement with Reverb Communications, a public relations (PR) firm. Reverb had several clients who published games in Apple's App Store. Reverb employees were found by the FTC to have written positive reviews. The FTC's position was that they were "shill" reviews; the term for this in the PR industry is "astroturfing," or creating fake grassroots messaging on behalf of a client. Reverb contended that the employees bought the games with their own money and wrote the reviews as individuals. Here are some of the reviews, as reported by the FTC:

- "Amazing new game."
- "ONE of the BEST."
- "[Developer of gaming application being reviewed] hits another home run with [gaming application being reviewed]."
- "Really Cool Game."
- "GREAT, family-friendly board game app."
- "One of the best apps just got better."
- "Developer of gaming application being reviewed] does it again!"(FTC Settling Case, n.d., a)

The complaint was settled by removal of the reviews from the App Store and the imposition of sanctions and reporting requirements on both the agency and the agency's owner, personally. Reverb owner Tracie Snitker said the following:

> The FTC has continuously made statements that the reviews are "fake reviews," something we question. . . . If a person plays the game and posts one review based on their own opinion about the game should that be constituted as "fake"? The FTC should evaluate if personal posts [from] these employees justifies this type of time, money, and investigation. It's become apparent to Reverb that this disagreement with the FTC is being used to communicate their new posting policy. (FTC Settling Case, n.d., a)

FTC attorney Stacey Ferguson told http://www.cnet.com the following:

> We're most concerned about the disclosure of the connection . . . so whether or not the employee actually did love the game or not, that wasn't really of consequence to us. We

> want them to disclose that they did have an affiliation with Reverb and the client when they're making those endorsements. (FTC Settling Case, n.d., b)

TechCrunch, a technology blog, performed an analysis of reviews by Reverb employees in the iTunes store. It found that several reviewers (a) only reviewed games whose developers were represented by Reverb, (b) only gave five-star (highest) reviews to those games, and (c) provided no disclosure of the material connection. TechCrunch also published a leaked client memo from Reverb that essentially bragged about this process:

> Reverb employs a small team of interns who are focused on managing online message boards, writing influential game reviews, and keeping a gauge on the online communities. Reverb uses the interns as a sounding board to understand the new mediums where consumers are learning about products, hearing about hot new games and listen to the thoughts of our targeted audience. Reverb will use these interns on Developer Y products to post game reviews (written by Reverb staff members) ensuring the majority of the reviews will have the key messaging and talking points developed by the Reverb PR/marketing team. (Cheating the App Store, n.d.)

TechCrunch concluded as follows:

> Ultimately, this is fraud. Plain and simple. Reverb Communications is using anonymized reviews as a way to boost sales, while lying to iTunes users. The worst part is many of these games stand by themselves. They have dozens of positive reviews from users (which we are assuming are not employees of Reverb). The developers are culprits as well.... We find it hard to believe they weren't privy to Reverb's actions. (Cheating the App Store, n.d.)

Note that the actions in question in this case actually predated the announcement of the revised endorsement guidelines. Furthermore, the settlement did not make mention of the individual employees/influencers who made the statements (aside from the owner) or of the brands and/or clients that enlisted Reverb's services to promote their products. The FTC focus is clearly on the agency. Employee and agency relationships must be disclosed in making product endorsements, reviewing products, and posting comments.

Legacy Learning Systems. The Gibson company, makers of the famous Les Paul electric guitar, lent its name to the DVD-and-booklet instruction series, "Gibson's Learn and Master the Guitar," published by Legacy Learning Systems. Legacy sold these DVDs online through a network of blogs and enthusiast websites. While the websites frequently offered rave reviews for the course, they did not disclose that the affiliate would earn up to 45% of the sale as a commission when a visitor purchased through a link. They were presented as reviews only.

According to the FTC, Legacy grossed more than $5 million through sales of this course (Firm to Pay FTC, n.d.). The FTC commissioners voted 5–0 to file a complaint against Legacy, which then paid a $250,000 settlement without admitting to guilt in the matter. David Vladeck, director of the FTC's Bureau of Consumer Protection, said the following:

> Whether they advertise directly or through affiliates, companies have an obligation to ensure that the advertising for their products is not deceptive. Advertisers using affiliate marketers to promote their products would be wise to put in place a reasonable monitoring program to verify that those affiliates follow the principles of truth in advertising. (Federal Trade Commission Settles, n.d.)

As part of the settlement, Legacy agreed to monitor two samples of affiliates for disclosure compliance—its top 50 affiliates and a second, random sample of 50 other affiliates—and submit monthly reports on its performance.

Here is the takeaway. A marketer must monitor affiliates and representatives who represent the brand or product and must remove and cut off payments to affiliates who do not comply with policies. The marketer has ultimate responsibility for the representations and misrepresentations of affiliates as well as for disclosures by affiliates.

50 Cent. On Saturday, January 8, 2011, rapper 50 Cent, born Curtis J. Jackson III, told his 3.8 million Twitter followers he was "lookin out for ya" by promoting an obscure penny stock. What he didn't tell his fans was that he held warrants in the company, which distributed "50 Cent" branded headphones and other products sold on infomercials (50 Cent Is Pumping, n.d.). During a single day he tweeted repeatedly about the stock and appeared on CNBC, increasing the stock's value 290% and raising the value of his 30 million shares to $8.7 million. Here is a sample tweet, which has since been deleted: "Ok ok ok my friends just told me stop tweeting about HNHI so we can get all the money. Hahaha check it out its the real deal."

While this event did garner some media attention, the FTC has taken no action to date. However, the tweets in question included no disclosure of his holdings and clearly got the attention of the FTC. Jackson's tweets might also have been of interest to the Securities and Exchange Commission (SEC) but seem to not have violated any existing guidelines. In fact, a 13D disclosure had been filed, indicating the holding to investors. However, individual investors would be unlikely to be aware of such disclosures, in particular in the Twitter environment within 140 characters. Had shares been sold within the given period, the SEC would likely have characterized this activity as an illegal "pump and dump" operation (50 Cent Is Pumping, n.d.).

Ashton Kutcher. Actor and entrepreneur Ashton Kutcher has expertly used social media to enhance his stardom and influence. According to http://twitaholic.

com, he is in the top 10 of Twitter users with more than eight million followers (Ashton Kutcher Twitter, n.d.). He uses his influence to promote social causes, his television show *Two and a Half Men,* and also companies in which he invests. In this last area, he rarely discloses his material connection. For example, on Sept. 6, 2011, Kutcher tweeted the following:

> Love that Conan O'Brien's Burbank TV Studio is up for rent on @Airbnb: 9/6/2011 1:02 PM. (Ashton Kutcher Could Face, n.d.)

Kutcher has a financial position in airbnb, among many other tech companies. He has shown a pattern of not disclosing his material connections in other social channels as well. In August 2011, he guest-edited "the social issue" of *Details* magazine. Among the tech and social media companies profiled in the issue, Kutcher is an investor in 12 of them. He offered no disclosure (Ashton Kutcher Could Face, n.d.).

Kutcher also shills for his companies in the mainstream media. In his second episode of "Two and a Half Men," which reaches up to 28 million viewers a week, he was shown using a laptop computer festooned with stickers for start-up tech companies. Gawker stated the following:

> Of the four monitor stickers we recognize, every single one belongs to a company in which Kutcher invests: iPad news aggregator Flipboard, travel searcher Hipmunk, textbook rental service Chegg, and check in service Foursquare, whose founder Dennis Crowley is giddy about the sitcom plug. (Here's the Ashton Kutcher Laptop, n.d.)

Following this incident, CBS ordered that the stickers be "greeked," meaning to have their identities obscured.

After the *Details* issue was published, Richard Cleland, assistant director of the division of advertising practices at the FTC, told *The New York Times* Bits blog that Kutcher could be disciplined for failing to disclose his material connections (Ashton Kutcher Could Face, n.d.). However, the FTC later posted to Twitter that Kutcher was not currently under investigation and that there were no plans to pursue the matter at that time.

Hyundai. In December 2011, the FTC decided to take no action on a blogger campaign developed by an ad agency on behalf of Hyundai automobiles. The unnamed agency recruited bloggers to build buzz over Hyundai's television ads appearing during Super Bowl XLV, providing gift certificates as an incentive. In its letter closing the investigation, the FTC said the following:

> First, it appears that Hyundai did not know in advance about use of these incentives, that a relatively small number of bloggers received the gift certificates, and that some

> of them did, in fact, disclose this information.
>
> Second, the actions with which we are most concerned here were taken not by Hyundai employees, but by an individual who was working for a media firm hired to conduct the blogging campaign. Although advertisers are legally responsible for the actions of those working directly or indirectly for them, the actions at issue here were contrary both to Hyundai's established social media policy, which calls for bloggers to disclose their receipt of compensation, and to the policies of the media firm in question. Moreover, upon learning of the misconduct, the media firm promptly took action to address it. (FTC Publishes Final Guides, n.d., b)

The FTC clarified the Hyundai case in a blog post written by Lesley Fair, dated December 22, 2011:

> So what does this mean for companies looking for more guidance on complying with the FTC's Endorsement Guides? M.M.M. OK, we just made up the mnemonic, but the principles are well established: 1) Mandate a disclosure policy that complies with the law; 2) Make sure people who work for you or with you know what the rules are; and 3) Monitor what they're doing on your behalf. (Using Social Media, n.d.)

The FTC's action in this case goes against many of its own principles. Hyundai appears to have been spared sanctions for these reasons: (a) the company was unaware of what the agency was doing on its behalf, (b) it took immediate action when informed of the promotion, and (c) the incentives were contrary to Hyundai's (and the agency's) published social media policies.

Summary of FTC Actions

Thus far, the FTC has fined only one company, and it has looked at only a handful of disclosure cases. There are a few takeaways from its action thus far:

- Only a few cases have been considered and only one fine levied.
- Advertisers must proactively teach influencers to disclose.
- Advertisers have a duty to monitor disclosure by influencers.
- Advertisers must be able to identify their influencers, document disclosure process and practices, and follow up with individuals to take down content or remove participants from future programs or compensation.
- Thus far, the FTC has largely used "raised eyebrow" techniques for enforcement.

Corruption of Online Discourse

While the FTC has looked at a handful of disclosure-related cases, nondisclosure and related kinds of corruption are rampant online. Popular tech writer John C. Dvorak said the following:

> It's a known fact that PR agencies and corporations have been burrowing into the community of online, "public" reviewers and obscure bloggers and easily corrupting them with trinkets. It's like a lost tribe being bribed with a pretty necklace of cheap polished rocks. Now there is some proof that there is a problem. (FTC Publishes Final Guides, n.d., b)

Many have observed that review sites may have been gamed by fake reviews. Arthur Frommer noted that hoteliers writing fake reviews of their own properties had infiltrated TripAdvisor reviews of Hawaii hotels (Ferreting Out Fake Reviews, n.d.). Many businesses have found Yelp reviews suspicious or arbitrary and then had difficulty getting fake reviews taken down once proof was offered (Yelp Refuses to Remove Fake Review, n.d.).

Two researchers at Cornell University examined reviewer behavior on Amazon Vine, a program for the most active reviewers at http://www.amazon.com (Cornell Chronicle, n.d.). The company noted the following:

> Vine helps our vendors generate awareness for new and pre-release products by connecting them with the voice of the Amazon community: our reviewers. Vine members, called Voices, may request free copies of items enrolled in the program and have the ability to share their opinions before these products become generally available. (Amazon.com Discussion Boards, n.d.)

The Cornell study found that 85% of Vine reviewers had received free products, and 78% "always or often" wrote a review when offered a free product. Thirty-five percent had noticed their words plagiarized in a review of another book or product. (Cornell Chronicle, n.d.)

A different sort of channel corruption comes from http://www.reputation.com, which, for a fee, offers to "Fix your Google results today." (Online Reputation Management Leader, n.d.). For example, the company's product, "Reputation Defender," "buries'negative and defamatory information on the Internet, bringing the stories you want to see to the top of your search results." (How to remove news articles from the web and protect your online reputation, n.d.) One product from http://www.reputation.com requires a medical patient to assign to his or her doctor the copyright to all consumer-generated reviews of that doctor as a condition of receiving treatment. This allows the doctor to simply delete any negative reviews on medical review sites (Doctors and Dentists, n.d.).

Another company that promises to help you manage your online reputation is Metal Rabbit Media. In addition to "search engine brand alignment," the company offers "Wikipedia Management":

> Wikipedia has quickly become one of the most dominant forces of search engine results. It consistently ranks in the top five for any brand or individual. Metal Rabbit Media has built a top-tier Wikipedia practice with a trusted track record of high profile edits. Having amassed a long history of engagement within the fickle Wikipedia community, we possess first-hand knowledge and expertise. Our Wikipedia services include content development, profile management, profile monitoring, and conflict resolution. (Metal Rabbit Media, n.d.)

Codes of Ethics Regarding Disclosure

While the public relations and marketing communities may be seen as contributing to increasing nondisclosure online, professional organizations in these fields consider disclosing material connections an important ethical responsibility. Of course, codes of ethics are a form of self-regulation, and many professionals do not belong to these organizations.

Here are some excerpts from codes of ethics that directly address disclosure.

Public Relations Society of America (PRSA). PRSA's code of ethics has a provision entitled "Disclosure of Information," based upon the principle that "open communication fosters informed decision making in a democratic society." The intent is "to build trust with the public by revealing all information needed for responsible decision making." (Public Relations Society of America Member Code of Ethics, n.d.)

The code states that a member shall do the following:

- Be honest and accurate in all communications.
- Act promptly to correct erroneous communications for which the member is responsible.
- Investigate the truthfulness and accuracy of information released on behalf of those represented.
- Reveal the sponsors for causes and interests represented.
- Disclose financial interest (such as stock ownership) in a client's organization.
- Avoid deceptive practices. (Preamble, n.d.)

Separately, PRSA has advocated for increased disclosure when there is an exchange between a PR practitioner and a journalist. Robert Frause, chair of the PRSA's Board of Ethics and Professional Standards, said the following:

> Readers, listeners, and viewers have the right to expect advance disclosure about anything that might compromise the integrity of the information they are getting. Journalists should be notified that any gift or in-kind service in exchange for placement should be clearly disclosed so the audience can make up its own mind about the information's value, bias, accuracy and usefulness. (PRSA Speaks Out, n.d.)

PRSA reiterated its stand on "Pay per Play" in its Professional Standards Advisory PS-9, dated October 2008. While it precedes the FTC ruling, the document supports disclosure (The New FTC Guidelines, n.d.).

PRSA convened a panel on disclosure in social media at its 2009 International Conference, held just a month after the FTC's guidelines took effect. Its online newsletter, PRSAY, offered a detailed summary of the FTC ruling on October 9, 2009, placing it squarely inside the existing code of ethics, summarizing as follows: "Essentially, the FTC applied longstanding principles to new media realities" (The New FTC Guidelines, n.d.).

International Association of Business Communicators (IABC). The IABC comprises membership largely working in PR and corporate communications. Its code of ethics covers much of the same ground as that of PRSA. Specific to the FTC ruling is Article 10: "Professional communicators do not accept undisclosed gifts or payments for professional services from anyone other than a client or employer" (IABC: Code of Ethics for Professional Communicators, n.d.).

IABC's magazine, *Communication World,* ran an article on the ruling in its January 2010 issue (member-only paywall). Additionally, IABC hosts a blog by Laura P. Thomas, who has written several posts on the ruling as it applies to so-called mommy blogs (LPT, 2009, November, n.d.).

Word of Mouth Marketing Association (WOMMA). The WOMMA has emerged as a leader of ethical behavior in the social space. The organization offers a detailed resource list on the FTC disclosure ruling as well as a code of ethics that includes specifics on disclosure (Your Single Source, n.d.).

The first four of eight standards of conduct reference some aspect of disclosure:

> Standard 1—Disclosure of identity: A WOMMA member shall require their representatives to make meaningful disclosures of their relationships or identities with consumers in relation to the marketing initiatives that could influence a consumer's purchasing decisions.

> Standard 2—Disclosure of consideration or compensation received: A WOMMA member shall require their representatives to disclose meaningfully and prominently all forms of consideration or compensation they received from the member, marketer or sponsor of the product or service. In other words, WOMMA members shall not engage in marketing practices where the marketer/sponsor or its representative provides goods, services, or compensation to the consumer (or communicator) as consideration for recommendations, reviews, or endorsements, unless full, meaningful, and prominent disclosure is provided.
>
> Standard 3—Disclosure of relationship: A WOMMA member shall require their representatives involved in a word of mouth initiative to disclose the material aspects of their commercial relationship with a marketer, including the specific type of any remuneration or consideration received.
>
> Standard 4—Compliance with FTC Guides: A WOMMA member shall comply with the Guides Concerning Use of Endorsements and Testimonials in Advertising promulgated by the Federal Trade Commission. See 16 C.F.R. §§ 255.0–255.5

WOMMA offers a comprehensive document with examples and best practices on how to disclose on blogs and in social media. It clearly describes good practice for disclosure in social media (Ethics, n.d.)

Good Practice Online—How to Disclose

For the independent blogger who wants to be in line with the FTC rules, disclosure is easy. If you received something of value, tell your readers. If in doubt, disclose anyway. But there are some other things you can do to fully inform your readers.

A site disclosure statement. In general, your disclosures should be presented with the relevant content. If you're reviewing a new automobile and a local dealer gave you the car to drive for 2 weeks for free, you should disclose that within your review. But you may find it useful to write a general disclosure statement as well. You can place this on its own page or on an "about" page. This is your chance to tell your readers—and potential sponsors or affiliates—about your personal "rules of the road," such as any long-term financial relationships you have. If you accept product or services for review, you can explain your process and what the sponsor can expect. For example, in this space you may reserve the right to write a negative review, or you can explain your policy for a sponsor's rebuttal of your review.

A disclosure statement doesn't need to be long or complicated. Remember, though, that this disclosure is in no way a substitute for disclosing with the relevant content.

The Shawn Collins Affiliate (tip) blog offers a concise policy on his "disclosure" page:

- I am never paid to do a review. I never accept money to review a product or service. I invest my own time to review and test products. I pay out of my own pocket the cost to produce all audio and/or video I record about products.
- If I create a link to a product or service in a review, sometimes I may get paid a commission if you purchase the product or service. These links are included after posts are written, and posts are never composed for the purpose of including advertising.
- No advertiser will ever influence the content, topics or posts made in this blog. (Disclosure Policy, n.d.)

Chris Brogan, blogger and social media marketer, aggressively monetizes his blog readership. Here are a few disclosures taken from an extensive list on his "about" page (excerpt):

> I write for American Express OPENForum.
>
> I am an advertising partner for Federated Media. (Check there if you want to place ads with me.)
>
> I am on the Advisory board for Hubspot.
>
> I am on the Advisory board for IZEA.
>
> I am cofounder of Third Tribe Marketing (affiliate link).
>
> I sometimes use Amazon Affiliate links to point to books I've reviewed.
>
> I'm an affiliate for Premise (affiliate link).
>
> I'm an affiliate for Screenflow (affiliate link)
>
> I'm an affiliate for Aweber email service.
>
> I'm an affiliate for Mark Dykeman's Unstuck (ebook).
>
> I use Skimlinks to monetize some product sales on this site. (About, n.d.)

Blog with integrity. At http://www.blogwithintegrity.com, a handful of bloggers approach self-regulation for the online community, offering a broad set of good practice standards. Bloggers may "sign the pledge" electronically and then display the "Blog with Integrity" badge on their websites as a seal of good practice.

Here's what the pledge says:

> By displaying the Blog with Integrity badge or signing the pledge, I assert that the trust of my readers and the blogging community is important to me.
>
> I treat others respectfully, attacking ideas and not people. I also welcome respectful disagreement with my own ideas.
>
> I believe in intellectual property rights, providing links, citing sources, and crediting inspiration where appropriate.

> I disclose my material relationships, policies and business practices. My readers will know the difference between editorial, advertorial, and advertising, should I choose to have it. If I do sponsored or paid posts, they are clearly marked.
>
> When collaborating with marketers and PR professionals, I handle myself professionally and abide by basic journalistic standards.
>
> I always present my honest opinions to the best of my ability.
>
> I own my words. Even if I occasionally have to eat them. (Blog with Integrity, n.d.)

Blog with Integrity was created by Susan Getgood, Liz Gumbinner, Kristen Chase, and Julie Marsh, each a blogger. It's unclear if Blog with Integrity is a non-profit organization, a business, or a service project which reflects the partners' personal interests.

CMP.LY. CMP.LY is an online service that explicitly addresses the FTC's material disclosure rules, offering a simple and direct way for a blogger to disclose specific kinds of relationships. CMP.LY is free to use, but the company offers a paid enterprise solution that also offers automated compliance monitoring, program reporting and disclosure analytics.

Because the CMP.LY URLs are short, they're appropriate for microchannels such as Twitter and Facebook. To use the service, simply select and post the tag that reflects your specific material connection:

> CMP.LY/0—No Material Connection
>
> CMP.LY/1—Review
>
> CMP.LY/2—Promo
>
> CMP.LY/3—Paid
>
> CMP.LY/4—Business
>
> CMP.LY/5—Affiliate
>
> CMP.LY/F—Financial Disclosures
>
> CMP.LY/H—Health/Pharmaceutical Disclosures
>
> CMP.LY/R—Rules/Contest/Promotion Disclosures
>
> CMP.LY/* is reserved for custom disclosures that don't fit the above categories. (CMP.LY, n.d.)

CMP.LY allows registered users to provide additional information to further describe the nature of the disclosed relationship. This enables a primary indication of the disclosure to be readable in a given post, tweet, or status update and an accompanying disclosure that provides complete information and context for that disclosed relationship.

The FTC has stated that simply putting a button on a post that says "disclosure" with a link to the disclosure policy isn't adequate (The FTC's Revised Endorsement Guides, n.d.). This suggests that CMP.LY is especially effective in microchannels, such as short message service (SMS or text) marketing or Twitter, where the compact disclosure would fit. For example, the link, http://CMP.LY/3/HPynQH, leads to a page with a photo and social links for Phil Campbell, along with this text:

> Full Disclosure Text
> I am a compensated member of the .tv Advocate program. As a .tv Advocate, I receive gifts, items of nominal value and cash payments for my content and participation in this program. This Disclosure applies to Verisign's .tv Advocate program. (Paid Post, n.d.)

Alternatives to CMP.LY for microchannels, also suggested by the FTC, are hashtags such as #ad, #paid, #spon, or #paid ad. The challenges with hashtags and other ad hoc solutions is that they do not provide the reader with context or information about the nature of the disclosed relationship, and they are difficult to monitor, in particular in the requirement to monitor influencers for the omission of a disclosure.

Criticism of the Disclosure Rules

In an open letter to the FTC, Interactive Advertising Bureau (IAB) President and CEO Randall Rothenberg acknowledged the FTC's role in regulating communication:

> All of us would agree that false and deceptive advertising should be stopped, and penalized when it slips through and is caught. We agree that paid testimonials and endorsements should be labeled. But in taking business ethics and attempting to give it the force of law, the Commission is stretching the definition of remuneration to ludicrous lengths. (Randall Rothenberg's Open Letter to FTC, n.d.)

But in a quote in an IAB press release, Rothenberg bristled at the notion that online media should be subject to these regulations while traditional media are not.

> What concerns us the most in these revisions is that the Internet, the cheapest, most widely accessible communications medium ever invented, would have less freedom than other media. These revisions are punitive to the online world and unfairly distinguish

> between the same speech, based on the medium in which it is delivered. The practices have long been afforded strong First Amendment protections in traditional media outlets, but the Commission is saying that the same speech deserves fewer Constitutional protections online. (IAB Calls on FTC, n.d.)

While the FTC denies that online publishers are second-class citizens in this regard, Rothenberg offered passages from the 81-page ruling that show otherwise (IAB Calls on FTC, n.d.). The FTC's interpretive guide contains softer language than the full document. For example, in the guide, the FTC states that its main interest is the advertiser's behavior, not the endorser's.

From a practitioner perspective, it's easy to support basic disclosure requirements. Rothenberg is taking a legal perspective to support long-term freedom for online publishers. To paraphrase Supreme Court Justice William O. Douglas, the FTC represents the nose of the camel inside the tent. Are the government's ends benevolent or sordid?

Historically, print media in the United States have enjoyed a high level of freedom relative to other media. In the online space, however, the distinction between journalist and "blogger journalist" is unclear. In virtually every category, an online "pure play" vies for audience with an established print-based alternative. Do you want to get your gossip from *People* magazine or http://perezhilton.com? Do you want to get automobile coverage from *Car and Driver* magazine or http://jalopnik.com? Do you want to get your news from *The New York Times* or *The Huffington Post?* Who is to say that the online competitors are not journalists? There is no license to be a journalist in the United States, no clear marker between journalist and nonjournalist.

Journalists Engaged in Commercial Speech

Because journalists, and print media in particular, have enjoyed high levels of press freedom, it's unlikely that the FTC will be investigating mainstream media anytime soon. This, despite the many perks enjoyed by the employees of daily newspapers, including review copies of books, free tickets, and preferential seating at concerts and sporting events. Readers accept these kinds of privileges because of the culture of mainstream journalism, which includes an ethos of independence and many layers of editing oversight designed to present an "objective" voice. But as you step away from newspapers and toward special interest publications, the notion of an industry captive becomes more plausible. Any magazine editor can share stories of endless swag, free events, and junkets designed to curry influence and favorable coverage.

Few journalists are willing to acknowledge this privileged relationship, which *Wall Street Journal* columnist Eric Felten broached in relation to his more highly regulated online colleagues.

"Jumbo are the shrimp and deep are the highballs at most media events," he wrote (Save Us, n.d.). His point was that the mainstream media get all sorts of free products—from books to reviews to the best seats at sporting events. Why should bloggers be held to a higher standard? Felten continued as follows:

> "I think this is absurd," says Alejandra Ramos, who writes a foodie blog called Always Order Dessert. She also happens to be an editor for a prominent women's fashion and lifestyle magazine and suggests that the FTC is laughably naive when it comes to the standards and practices of her business: "Magazines are sent free products all the time." So much so that staffers have to be encouraged to take products home just to clear out the "beauty closets." And yet, Ms. Ramos says, when a big glossy does a feature on "seven mascaras that will make your lashes look longer, they do not appear with a disclaimer that 'L'Oreal sent us this mascara for free.'" Why, she asks, should the law treat bloggers any differently? (Save Us, n.d.)

Scope of the Blogosphere

The FTC has been charged with patrolling a very large frontier, populated with millions of blogs and billions of status updates, tweets, and other ephemera. We have seen that the traditional communicators—journalists, marketers, and PR professionals—are broadly bound by their respective cultures, economic models, and codes of ethics. Generally speaking, they disclose appropriately.

As you step away from the mainstream media, however, things get murky. Now that everyone can publish, we're seeing a broad range of content, approaches, and financial models. How can you trust a publisher you don't know? Ultimately, it's every person's duty to follow the links and vet the information sources he or she relies upon. Media literacy training can help, as can a trusted content curator.

But overall, the FTC can't solve the disclosure problem. The agency has stated rules for good practice. Through rulings and statements, it can articulate exactly what the rules mean. Over time, this process should influence the practice of mainstream communicators. But, like an iceberg in the North Atlantic, most of the threat will remain unseen by the agency and thus invisible to consumers.

References

50 Cent Is Pumping a Penny Stock on Twitter, and Its Shares Are Going Nuts. (n.d.). Retrieved from http://www.businessinsider.com/50-cent-hnhi-2011–1

About. (n.d.). Retrieved from http://www.chrisbrogan.com/about/

Amazon.com Discussion Boards #26876.1. (n.d.). Retrieved from http://forums.prosperotechnologies.com/n/mb/message.asp?webtag=am-custreview&msg=26876.1&&liaagc=y&redirCnt=1&lgnJR=1

Ashton Kutcher Could Face Questions About Disclosure—NYTimes.com. (n.d.). Retrieved from http://bits.blogs.nytimes.com/2011/08/18/ashton-kutcher-could-face-questions-about-disclosure/

Ashton Kutcher Twitter Stats & Rankings (aplusk) | Twitaholic.com. (n.d.). Retrieved from http://twitaholic.com/aplusk/

Blog with Integrity, Inc. (n.d.). Retrieved from http://www.blogwithintegrity.com/

Cheating the App Store: PR Firm Has Interns Post Positive Reviews for Clients [UPDATED] | TechCrunch. (n.d.). Retrieved from http://techcrunch.com/2009/08/22/cheating-the-app-store-pr-firm-has-interns-post-positive-reviews-for-clients/

CMP.LY. (n.d.). Retrieved from http://cmp.ly/

Cornell Chronicle: Study Hones in on Amazon Reviewers. (n.d.). Retrieved from http://www.news.cornell.edu/stories/June11/PinchAmazon.html

Disclosure Policy. (n.d.). Retrieved from http://blog.affiliatetip.com/affiliate-link-disclosure/

Doctors and Dentists Tell Patients, "All Your Reviews Are Belong to Us." (n.d.). Retrieved from http://arstechnica.com/tech-policy/news/2011/05/all-your-reviews-are-belong-to-us-medical-justice-vs-patient-free-speech.ars

Ethics. (n.d.). Retrieved from http://womma.org/ethics/code/

Fashion Bloggers Run Afoul of New FTC Rules? (n.d.). Retrieved from http://jezebel.com/5463427/fashion-bloggers-run-afoul-of-new-ftc-rules

Federal Trade Commission—About Us. (n.d.). Retrieved from http://www.ftc.gov/ftc/about.shtm

Federal Trade Commission Settles Against Legacy Learning Systems for Misleading Endorsements Made by Affiliate Ad Nauseum. (n.d.). Retrieved from http://www.adnauseumblog.org/federal-trade-commission-settles-against-legacy-learning-systems-for-misleading-endorsements-made-by-affiliate/

Ferreting Out Fake Reviews Online—NYTimes.com. (n.d.). Retrieved from http://www.nytimes.com/2011/08/20/technology/finding-fake-reviews-online.html?_r=1

Firm to Pay FTC $250,000 to Settle Charges That It Used Misleading Online "Consumer" and "Independent" Reviews. (n.d.). Retrieved from http://www.ftc.gov/opa/2011/03/legacy.shtm

FTC Gives Ann Taylor a Pass in First "Blog Disclosure" Investigation | Techdirt. (n.d.). Retrieved from http://www.techdirt.com/articles/20100514/0111169422.shtml

FTC Publishes Final Guides Governing Endorsements, Testimonials. (n.d., a). Retrieved from http://www.ftc.gov/opa/2009/10/endortest.shtm

FTC Publishes Final Guides Governing Endorsements, Testimonials. (n.d., b). Retrieved from http://www.ftc.gov/opa/2009/10/endortest.shtm

FTC Responds to Blogger Fears: "That $11,000 Fine Is Not True" | Fast Company. (n.d.). Retrieved from http://www.fastcompany.com/blog/jennifer-vilaga/slipstream/ftc-bloggers-its-not-medium-its-message-0

FTC Settling Case over "Fake" iTunes Reviews | Politics and Law—CNET News. (n.d., a). Retrieved from http://news.cnet.com/8301–13578_3–20014887–38.html

Heinrichs, J. (2007). *Thank you for arguing: What Aristotle, Lincoln and Homer Simpson can teach us about the art of persuasion.* New York, NY: Three Rivers Press.

Here's the Ashton Kutcher Laptop That CBS Banned. (n.d.). Retrieved from http://gawker.com/5844477/heres-the-ashton-kutcher-laptop-banned-by-cbs

How to remove news articles (n.d.) Retrieved from http://www.reputation.com/reputationwatch/articles/how-remove-news-articles-web-and-protect-your-online-reputation

IAB Calls on FTC to Rescind Blogger Rules; Questions Constitutionality. (n.d.). Retrieved from http://www.iab.net/about_the_iab/recent_press_releases/press_release_archive/press_release/pr-101509

IABC: Code of Ethics for Professional Communicators. (n.d.). Retrieved from http://www.iabc.com/about/code.htm

LPT, 2009, November. (n.d.). Retrieved from http://laurapthomas.x.iabc.com/2009/11/

Metal Rabbit Media—Brand Your Web. (n.d.). Retrieved from http://metalrabbitmedia.com/

New F.T.C. Rule Has Bloggers and Twitterers Mulling—NYTimes.com. (n.d.). Retrieved from http://www.nytimes.com/2009/10/15/fashion/15bloggers.htm

Online Reputation Management Leader: Reputation.com. (n.d.). Retrieved from http://www.reputation.com/

Paid Post | CMP.LY. (n.d.). Retrieved from http://cmp.ly/3/HPynQH

Payola and Sponsorship Identification. (n.d.). Retrieved from http://transition.fcc.gov/eb/broadcast/sponsid.html

Preamble. (n.d.). Retrieved from http://www.prsa.org/AboutPRSA/Ethics/CodeEnglish/

PRSA Speaks Out on "Pay for Play," Strengthens Code of Ethics' Transparency Provisions | PRSA Newsroom. (n.d.). Retrieved from http://media.prsa.org/article_display.cfm?article_id=1297

Public Relations Society of America Member Code of Ethics (n.d.) Retrieved from http://www.prsa.org/aboutprsa/ethics/codeenglish/

Randall Rothenberg's Open Letter to FTC. (n.d.). Retrieved from http://www.iab.net/public_policy/openletter-ftc

Reputation Defender (n.d.) Retrieved from https://www.reputation.com/red

Save Us From the Swag-Takers—WSJ.com. (n.d.). Retrieved from http://online.wsj.com/article/SB10001424052748703746604574461220828153720.html

The FTC's Revised Endorsement Guides: What People Are Asking | BCP Business Center. (n.d.). Retrieved from http://business.ftc.gov/documents/bus71-ftcs-revised-endorsement-guideswhat-people-are-asking

The New FTC Guidelines: Cutting through the Clutter | PRSAY—What Do You Have to Say? (n.d.). Retrieved from http://prsay.prsa.org/index.php/2009/10/09/the-new-ftc-guidelines-cutting-through-the-clutter/Guides concerning use of endorsements and testimonials in advertising. 16 C.F.R. §§ 255.0–255.5. (2009). Retrieved at http://ecfr.gpoaccess.gov/cgi/t/text/text-idx?c=ecfr&sid=38ef2be645af7a656be40e21324801c9&tpl=/ecfrbrowse/Title16/16cfr255_main_02.tpl

Using Social Media in Your Marketing? Staff Closing Letter Is Worth a Read | BCP Business Center. (n.d.). Retrieved from http://business.ftc.gov/blog/2011/12/using-social-media-your-marketing-staff-closing-letter-worth-read

Yelp Refuses to Remove Fake Review of Anella—Yelp Wanted—Eater NY. (n.d.). Retrieved from http://ny.eater.com/archives/2011/07/yelp_refuses_to_remove_fake_review_of_anella.php

Your Single Source for All Things FTC Related. (n.d.). Retrieved from http://womma.org/ftc/

CHAPTER FIVE

Ethics of Citizen Journalism Sites

Jessica Roberts & Linda Steiner

Now widely equipped with tools for recording and sharing content online, citizens have become much more active in producing and distributing news. The resulting expansion of the network of "mass self-communicators" (Castells, 2007) has led to a wealth of news and information content online that comes from outside the walls, or firewalls, of professional journalism organizations. Citizens not otherwise employed as journalists find themselves with access to tools for recording and sharing text, photos, video, audio, and other content more quickly and easily than ever before. Apps for digital cameras contained within cell phones enable users to upload pictures instantly to social networks, blogs, and photo-sharing sites. The same software—FinalCutPro—is used to edit feature-length films and YouTube videos. Not only are the costs going down but some new technologies start out at prices affordable to citizens. And increasingly, new and coming technologies will have ethical implications. New software for cell phones enables users to instantly obtain personal and professional information about people who are nearby, along with precise geolocation and photographs. StreetSpark uses geolocation to find and introduce users nearby with similar "likes." That is, these technologies enable not only vertical but also lateral surveillance as well as broad, instant, and seemingly "unmediated" dissemination. Promoted as dating tools, these technologies will undoubtedly be used by enterprising journalists, including citizens not operating within editorial hierarchies. The technologies reshape journalism—redefining journalism, how it is done, and, of course, who is a journalist—and raise new ethical challenges.

Already, both independent citizen news websites and professional news outlets allow and even encourage users to contribute reports, photos, comments, and/or their "likes." Sites where citizens contribute journalistic content vary greatly in their structures, goals, economic models and, especially, in their ethical principles or guidelines. Platforms for sharing information often include no more than minimal rules, such as requiring users to provide a username and password. Most of the 34 sites we looked at state general values or provide instructions for contributors about how to post information. On the other hand, few offered guidelines for gathering information; instead, they merely provide an end user license agreement or terms of use policy outlining the liability of each party—generally, limiting the liability of the owners and managers of the site. Moreover, although users must click a button stating that they have read the policy, the policy's placement implies that users are not actually expected to read it; indeed, the detailed, formal legalistic language actively discourages reading. The same news organizations that hire ombudsmen and publish lengthy ethics statements offer little or no ethics guidance to commentators; they simply ban obscenity and rude behavior. Most conventional news organizations borrow, albeit inconsistently, from their parent outlet in offering thin rules for content contributed by nonprofessionals.

Sites also vary in the extent to which they attempt to prevent or punish violations of the codes or guidelines. Some, for example, merely follow Clay Shirky's (2008) "publish, then filter" model: They allow participants to complain and to insist that content be deleted. Meanwhile, the damage itself may be irreparable. Even the most sophisticated of these citizen news sites have given far less attention to ethical conflicts and dilemmas than have many online crowd-sourced projects. In addition, there has been no scholarly attention to this question.

This research analyzes what is said about ethics by 34 citizen journalism sites, including some sponsored by commercial print and broadcast news media outlets, as well as those that are largely self-organized (see Appendix 1 for a list of the sites of various types). We particularly looked for statements on these sites about ethical conflicts, how they justify their ethical principles or explain why they ignore ethics. We hope the analysis has practical value for sites trying to articulate useable ethical principles; Appendix 2 lists what we consider best practices.

Citizen Journalism

Australian new media scholar Axel Bruns (2008) dates citizen journalism to proactive and highly networked organizing by Indymedia activists before the 1999 World Trade Organization meeting in Seattle. The concept almost immediately spread globally. Using the slogan "Every citizen is a reporter," the pioneering South Korean site OhmyNews launched in 2000; by 2009, some 70,000 citizens from around the world had written stories for the site's volunteer editors (Woyke, 2009).[1] Jay Rosen (2006) was an early advocate of journalistic involvement by citizens—"the people formerly known as the audience." Rosen famously asserted on his Pressthink blog that news media no longer run "one-way," with a few news organizations dominating the discourse while everyone else listens in isolation. Now, according to Rosen, anyone who wants to can be a journalist: The "horizontal flow, citizen-to-citizen, is as real and consequential as the vertical one" (para. 10).

A former journalist for several conventional news organizations and now running the Knight Center for Digital Media Entrepreneurship at Arizona State, Dan Gillmor (2004) also popularized the concept. Other advocates have celebrated how citizens can help produce information. The term "produser" was coined by Axel Bruns to describe an individual who both uses and produces content online. Describing how ordinary people could bear witness to the London bombings of 2005, Stuart Allan (2006) wrote: "The familiar dynamics of top-down, one-way message distribution associated with the mass media are being effectively, albeit unevenly, pluralized.... Online news is an increasingly collaborative endeavour, engendering a heightened sense of locality, yet one that is relayed around the globe in a near-instant" (pp. 2, 19). In *We're All Journalists Now,* Scott E. Gant (2007) explained:

> Much of what is worth knowing, and worth thinking about, is neglected by the mainstream media. Now, with the rise of citizen journalism, many more people are passing on their observations and ideas, playing a role previously occupied only by members of the institutional press. (p. 45)

Of course, citizens are involved in creating content online in very different ways, ranging from merely commenting on or sharing a news story written by a professional journalist to creating and uploading entirely original reports in video or text form. Many different forms of participation are often thrown together under the label "citizen journalism," despite varying levels of engagement and journalistic intention on the part of participants, as well as different funding models and geographic domains. For example, of the sites we examined, some are highly—hyper—local, while others are international on principle. Jan Schaffer (2007) lumped together as citizen journalism a variety of forms: "hyperlocal citizen media" sites entirely run by volunteers; sites owned and controlled by legacy organizations; hybrids of citizens supervised and/or trained by paid staffers; totally independent yet professional journalists incorporating citizen-written material, sometimes with foundation support or investors; profit and nonprofit sites begun by individuals; and community cooperatives where volunteers share decision-making, sometimes at formal meetings. In recent years, Twitter and a variety of other social media activities have also been included under the umbrella of citizen journalism.

Crowd-sourced models and the terminology of collaboration have also been applied to citizen journalism, often by those celebrating citizens' participation. James Surowiecki (2004) argued that larger, more diverse groups could often result in better decisions and higher quality information than smaller, elite groups. His theory of the "wisdom of crowds" is that "the best collective decisions are the product of disagreement and contest, not consensus and compromise" (p. xix), much as John Stuart Mill's (1869) theories about free expression proposed that allowing all opinions to be aired would ultimately result in better opinions. This perspective was further developed by Bruns (2008), who argued that the mass of individuals online can effectively produce and filter content. Shirky (2008) likewise agreed that the collective of individuals online can publish information and then filter through it rather than rely on legacy news institutions acting as gatekeepers.

Despite enthusiasm about the presumed "outsider" perspectives of citizens widening the scope of information published, citizen journalists deploy the same techniques for finding information and types of sources as their professional counterparts. Citizens and professionals do, however, differ in the extent to which they rely on various kinds of sources (Reich, 2008). Schaffer (2007) suggested that citizen journalism formed a bridge between traditional forms of journalism and classic civic participation. Sue Robinson (2011) sees citizen contributions (commenting, sharing or otherwise responding to content) as part of "journalism as process." This understands journalism not as the output of journalists employed at legacy news organizations but as an ongoing process through which versions of events slowly evolve.

Critics, including editors, reporters, and media critics (see Dowd, 2009; Skube, 2007), point to a lack of qualification and editorial oversight, sometimes mocking the image of a "guy in pajamas" pontificating about news events. Others cite instances of misinformation that originated from citizens sharing information online. Nicholas Lemann (2006) argued that citizens cannot do the work needed to monitor institutions of authority or "mount the collective challenge to power which the traditional media are supposedly too timid to take up." Others worry that the enthusiasm over citizen involvement in journalism understates the degree of citizen passivity (especially if this is considered in international terms) and/or ignores the continuing power of media industries. Sasha Costanza-Chock (2008) criticized user generated content as free cultural product for monetization and cross-licensing; "'participation' means free user data to mine and sell to advertisers, and all user activity is subject to surveillance and censorship" (p. 857). Former newspaper editor Tom Stites criticized citizen journalism as mostly the province of "a rather narrow and very privileged slice of the polity—those who are educated enough to take part in the wired conversation, who have the technical skills, and who are affluent enough to have the time and equipment" (quoted in Gillmor, 2004, p. xxix). José van Dijck (2009) suggested that an emerging rule of thumb is that only 1 in 100 people will be active online content producers; 10 will comment, and the remaining 89 will simply view.

What Can Citizen Journalists Do?

Not all citizen "produsers" are trying to be journalists. Although 78% of respondents to Schaffer's (2007) survey said they provide "journalism" and 46% said they provided mainly news and information, citizens do not consistently call themselves journalists when involved in creating news content (p. 23). They may have different motives altogether. Creating content may help people feel important and connected to others (Daugherty, Eastin, & Bright, 2008). Research on Wikipedia suggests that contributors to the online encyclopedia are strongly motivated by being part of a community and making contributions to society, enhancing self-efficacy and self-confidence, learning things and developing skills, partaking in intrinsic pleasure and fun, sharing with others, holding to political principles, identifying with Wikipedia's values, and being challenged intellectually (Johnson, 2008; Nov, 2007; Yang & Lai, 2010). Citizen journalists are likewise probably motivated at least in part by non-news values.

Regarding the question of whether citizen journalists can replace professionals, the blogosphere tends to resoundingly answer "no." On the other hand, the failure of news media's financial/business model has precipitated extended discussions of the role of amateurs and audiences. CNN explicitly connected its November 2011 layoff of photojournalists and other professionals to consumers and prosumers with increasingly accessible but high quality technologies. In 2008 the publisher of the online daily, *Pasadena Now,* fired all seven staff members, and, using Craigslist, outsourced news and features to freelancers in India who are paid by the word. One such freelancer told syndicated columnist Maureen Dowd (2008), "I try to do my best, which need not necessarily be correct always." Dean Singleton, chairman of the Associated Press and head of the MediaNews group, announced that his company is considering similar outsourcing; most of the preproduction work for the group's California papers was already outsourced to India, cutting costs by 65%.

Journalism Ethics Off-Line and Online

Professional journalists as individuals are assumed to have ethical responsibilities and commitments by virtue of being professionals. More importantly, journalists (at least in democratic societies) have a special responsibility to provide the information citizens need to make political decisions. This theory of a press responsible to the public interest is more and less consistently laid out in many normative documents, from The Hutchins Commission on the Freedom of the Press (1947) to the Knight Commission on the Information Needs of Communities in a Democracy (2009).

A broad range of journalism institutions, including newspapers, professional associations, and journalism schools, promote journalism ethics. Unlike in professions such as law and medicine, of course, journalists are not required to have professional education, to pass tests, or to be certified. Many major newspapers refuse to establish an official code of ethics, presumably to protect themselves legally. (The First Amendment can be invoked here.) Nonetheless, as a condition of accreditation, journalism schools must require students to study journalism ethics. Moreover, many news media and journalism organizations do have codes of ethics or explicitly state ethical goals, values, or missions. Generally, accuracy and attribution are considered paramount, as is the duty to minimize harm. The fact that ethics codes are commonly framed from a social responsibility perspective explains why codes are generally perceived to lie within the domain of ethics and moral philosophy as opposed to law and precedent; moreover, codes articulate both a set of rules for "normal journalistic practice" as well as to inspire the highest ideals in the profession (Brennen & Wilkins, 2004).

Online contexts may require adjustments in the behaviors and practices of journalists, professional or not. Cecilia Friend (2007a) noted that accuracy, verification, and source transparency demand additional attention when journalists assess and use information from online sources. Journalists also face new difficulties in separating editorial and commercial content in a web environment combining proprietary information, sponsored content, and advertisers (Friend, 2007b). Indeed, the visual identification of analysis and commentary as separate from news that appears in a printed newspaper may be lost in its online version.

According to Deborah Johnson (1997), the scope, anonymity, and reproducibility of online communication raise new ethical problems and thus may demand special consideration by people sharing information online. A network, she said, magnifies the impact of an act many times over; one moral implication of scope, or power, is that individuals engaged in powerful activities are expected to take greater care. Second, although off-line anonymity requires effort on the part of someone seeking anonymity, anonymity is often the natural state online. Third, "Information can be reproduced online without loss of value and in such a way that the originator or holder of the information would not notice" (Johnson, 1997, p. 62).

Citizen Journalists and Ethics

A few codes of ethics developed for online journalism have a great deal in common with professional journalists' codes of ethics. The three principles of Cyberjournalist.net's (2003) Blogger's Code of Ethics are to be honest and fair, accountable, and to minimize harm. Martin Kuhn (2007) proposed a mix of ethical and practical guidelines for bloggers: promote interactivity, free expression, and the "human" element, and strive for factual truth and transparency. An ethics code for online journalism developed at the University of Southern California mandates honesty and disclosure of sources, conflicts and motivation to publish.[2]

Citizen journalists not only lack professional status but their emergence is relatively recent. So to judge them by different, even lower, standards would not be surprising. Lee Salter (2009) proposed understanding citizen journalism in relation to law, although applying law is especially challenging in an online environment said to transcend jurisdictions. For example, he pointed out that citizen journalists may seek rights as journalists but simultaneously seek protection in anonymity and a sense of freedom stemming from exploiting the supposed deterritorialization provided by the Internet. But too often, however, the hype about citizen journalism ignores questions about responsibilities of citizen journalists. The implication is that citizen participation is good regardless of whether it's ethical. Moreover, the failure to address citizen ethics is at odds with the logic that citizen journalism is an important phenomenon—an argument asserted by both its celebrators and its critics. The Knight Citizen News Network[3] provides a list of principles—accuracy, thoroughness, fairness, transparency, and independence—that offers a glimpse of what sorts of ideas might deserve the attention of citizen journalists and organizations seeking their input. The *Huffington Post*, which has enlisted the help of readers to generate election coverage "off the bus," urges citizen journalism to conform to specific standards: stick to facts, avoid hearsay, omit irrelevant opinion, never plagiarize, never edit or alter photos, identify yourself when reporting, identify and fact-check sources. These express no ethical principles to which citizens might aspire, although they articulate rules governing both the content of citizen journalism and the behavior of citizen journalists.

Methods

In the interests of maximum diversity, we purposively selected at least a few examples in each category of online sites that we could identity that might prominently feature, if not exclusively offer, citizen journalism (see Appendix 1 for a complete list). Sites that aggregate, publish, or allow citizen contributions vary in their structure: Some solely collect reportage and/or commentary of citizens and provide no editorial oversight. Others are essentially legacy journalism organizations that carefully select and edit contributions from citizens. Still others function as a user-moderated or user-edited site in which users entirely control content and editing. Some offer a great degree of editorial control over content published on the site; in contrast, content-sharing sites allow unfettered posting of just about anything. Our sample included sites that collect citizen news for possible pay, hybrid and often hyper-local sites that accept selected citizen contributions with editorial oversight and also provide news produced by staff journalists and corporate-owned sites that allow citizen contributions ranging from comments on news stories to full reports with video. Some had a global reach; others were highly local. Some expressly repudiated the conventions of mainstream journalism. In contrast, some embraced this, including Third Report, whose term "Third Men and Women" refers, as if these terms are clear to everyone, to "traditional" reporters writing from the neutral perspective of a third person—fair and unbiased. Four sites were attached to university-level journalism schools. Altogether we identified 39 possibilities, but five were discarded as either no longer operational or insufficient to "qualify" as citizen journalism.

In analyzing the sites themselves, we clicked through the entire site or as much as possible. We took notes on what appeared on the home page and elsewhere (and where specifically on any page relevant material appeared). Background dimensions of interest included who seemed to own or run the site, what the site announced as its purpose, when the site was established, and what was necessary for individuals to post text, photography, video, or audio. The centerpiece of the analysis involved asking a number of questions about ethics: What kinds of codes were offered, if any? Did the ethics discussion go beyond forbidding clearly taboo behaviors that would be off limits for everyone? Did its principles (if, indeed, ethical practice was cast in terms of principles) address a range of quandaries and dilemmas that professional journalists confront, such as those having to do with protecting (or revealing) a source's identity or the use of deceptive or other surreptitious methods of gathering information? Were conflicts of interest (real or perceived) or checkbook journalism mentioned? Were ethical dilemmas involving photos, graphics, video, audio, and/or sound bites mentioned, including whether "manipulation" might, or might not, be appropriate? What was the role of taste or good taste? Furthermore, why were citizen journalists held to ethical standards? On whose behalf were citizens directed to be accountable (i.e., only each other or also sources or audiences)? What kinds of enforcement mechanisms were established? Were citizen journalists encouraged to admit their own ethical violations or those of others? Were the ethics merely technical and procedural, or did they speak to a larger mission and responsibility of journalism? If readers, fellow citizen journalists, or editors evaluated the articles posted, on what basis was this evaluation made (i.e., was it a matter of popularity)? What adjectives were applied to citizen journalism? How was diversity understood, if it was mentioned? Finally, did the site provide resources, citations, or literature for further thinking about ethics? Did the site link to organizations specializing in ethics[4] or codes of ethics such as those of the Society of Professional Journalists (SPJ), the Radio-Television News Directors Association, the National Press Photographers Association, or even codes in other professions? Was ethics contextualized in terms of theories? Were non-U.S. resources and principles highlighted?

Initially, both authors analyzed six particular sites, to check for consistency and agreement in analysis. The other sites were analyzed separately, using qualitative methods; at the end, we double-checked by returning to the sites. Some of the features were listed in a spreadsheet so that we could "count" relatively common features. Our goal was not to establish some artificially high standard for these sites or to demonize sites for failure to fully answer each of the criteria mentioned earlier. Rather, we were trying to inventory ways in which these sites introduce education, standards, and resources about ethics so that citizen journalists could develop a serious ethical awareness appropriate to the serious aims claimed on their behalf.

Findings

Of the three sites that dealt extensively with ethics, the best example is Twin Cities Daily Planet, a hyper-local site serving Minneapolis and St. Paul. The Daily Planet publishes original articles and blog entries and also republishes content from community media partners. The Daily Planet's attention to ethics is not surprising. The Daily Planet is a project of the Twin Cities Media Alliance, a nonprofit organization bringing together professionals and engaged citizens to improve the quality, accountability, and diversity of the local media; its executive director is Jeremy Iggers (1998), a former *Minneapolis Star Tribune* writer whose doctoral dissertation addressed journalism ethics. In addition to offering extensive editorial tips, tools, and guidelines for writers, the Daily Planet sponsors writers' groups, two "courses" in citizen journalism, workshops, and a weekly "User Training, Open Newsroom and Happy Hour." Furthermore, the site suggests that journalism is not simply a matter of uploading any interesting content, as many sites do. The Daily Planet asserted: "Journalism is a craft. Doing it well takes skills and lots of practice. But we believe anyone can learn the tools it takes to report and write a story." Calling journalism a craft rather than a profession is arguable, but at least this site emphasizes the effort and time needed to do journalism well.

A small India-based site launched in 2006 requires all its "citizen reporters" to "abide by a strict Code of Ethics." MyNews Reporter's Code of Ethics requires its citizen reporters to identify themselves as citizen reporters while covering stories and to use "legitimate" methods to gather information. The rest of the principles were stated in the negative: The ethical citizen reporter does NOT spread information that is false, distorted, or based on groundless assumptions; use abusive, vulgar, or otherwise offensive language; damage reputations of people or infringe on personal privacy; or seek personal profit. Moreover, in an unusual addition, the code described citizen reporters as apologizing fully for incorrect or inappropriate coverage.

Wikinews also has an elaborate discussion of conduct—not a surprise, given the institutional backing of the Wikimedia Foundation and its 10 years of experience in dealing with controversial but collaborative production and collective decision making:

> Articles are written collaboratively for a global audience. We strive at all times to meet the policy of using neutral point of view, ensuring our reporting is as fair as possible. Furthermore, everything we write is cited, to maintain the highest standards of reliability.

The terms were never problematized: Wikinews never explained why neutrality is required, how it ensures fairness, or how citations ensure reliability. Nonetheless, the site repeated these standards across several sections and explained the procedures for reporting abuses and appealing negative decisions.

Most other sites gave ethics little attention and rarely took seriously the ethical obligations of citizen journalists, beyond recognition of issues relating to privacy, accuracy, and copyright, as well as spamming. Generally the emphasis was on prohibited behaviors, such as not allowing content that is abusive, rude, inappropriate, obscene, or racist. Most sites made clear that engaging in specifically prohibited behaviors could result in removal of content or banning a user from the site. Prohibited behaviors were merely listed in legalistic language, apparently intended to eliminate liability for the site's owners; most commonly banned were copyright violations, harassment, illegal or unlawful acts, and junk mail, spam, and advertising. Nearly all sites provided a Terms of Service or User Agreement, to which users indicated agreement by clicking a button on the registration page.

A few sites discussed libel and defamation in a more general context than the terms of use policy, although very few explained what might constitute libelous or defamatory content. The Canada-based technology site, DigitalJournal.com, for example, claimed that its 32,000 citizens and digital journalists work "24–7" to report news from 200 countries. DigitalJournal's extensive discussion of Canadian libel law cautioned citizen journalists to be "libel-conscious" by getting all facts right and backing them up with evidence. The company boasted of its accuracy, research, first-hand reporting, and its requirement that stories have three to five sources. "This rule will make you a true representative of today's social media environment and link economy, and it will open your mind to other information that can help strengthen your original idea." But if getting hit with a lawsuit means "a world of pain," to quote DigitalJournal.com, no reference was made to being accused of ethical violations.

Some sites heralded their openness and weak (not their word) editing process. "Your story will never be edited as to content or substance," iBrattleboro announced, although citizen reports might be editing for spelling, punctuation, and/or length. Other sites substituted algorithmic approaches to quality for ethics. Newsvine was one of several to brag that no editors had control, because "you decide what appears here." At Newsvine, readers could click the "vote" button; the story's resulting popularity score dictates what appears on the site's front page. The lack of editorial oversight was sometimes contrasted by a strong community of users who were expected to accomplish much of the editing or moderating.

The news-sharing or news aggregator sites, including Digg, Reddit, Slashdot, and StumbleUpon, rely on users to moderate each other and build reputation through peer evaluation of behavior on the site. With their well-developed and extensive community guidelines, these sites largely succeed in maintaining a strong, engaged community of users committed to the site's quality. They moderate news stories and each other, without interference or oversight from an editorial or supervisorial staff. While the sites do not address what might constitute ethics for posting, they leave it up to users to determine what content they consider useful, relevant, and important. For example, Reddit's FAQ explains why it has moderators: "These communities distinguish themselves through their policies: what's on- and off-topic there, whether people are expected to behave civilly or can feel free to be brutal, etc. The problem is that casual, new, or transient visitors to a particular community don't always know the rules that tie it together." The site's "Reddiquette" (an informal expression of Reddit's community values) includes instructions to keep submission titles factual and opinion-free, moderate based on quality rather than opinion, and "actually read an article before you vote on it." In explaining the site's "karma" scores (based on up or down votes from other users), the following is offered to explain why users might want to accumulate karma: "Don't set out to accumulate karma; just set out to be a good person, and let your karma simply be a reminder of your legacy." The ethics of these sites are otherwise not outlined in explicit terms but generated collectively by users through their voting and moderating behaviors. Because these sites aggregate news stories but generally do not accept original stories, some of the ethical concerns for citizen journalists who create original news stories are irrelevant here—especially those relating to ethical conduct in news gathering. However, admonitions about submitting false or misleading stories, or even stories that have already been submitted by another user, are common on these sites.

A more complicated algorithm was developed by Allvoices, a self-described "global, open-media news site where anyone can report from anywhere." "We envision a true 'people's media,' where each report generates multiple points of view and a direct, emotional connection to what really matters." A proprietary five-point scale, "Report Credibility Rating System"—its claim to fame—indicates the trustworthiness and relevance of a citizen reporter's story, based on community rankings of the content's "buzz" and the author's reputation. Allvoices said:

> The Report Credibility rating scale is determined by the level of community interaction and response to the story, reporter reputation over time, and the power of the Allvoices intelligent news analysis platform. . . . As the community interacts with the report and more confirming evidence is discovered, the report's credibility will change.

That is, the site expressly allows dissemination of stories it has not vetted and written by people who have not yet earned credibility and then makes reputation a matter of branding and marketing, not democracy.

Among the sites least concerned with ethics were the ones aggressively but also vaguely (i.e., never mentioning real dollar amounts paid to reporters) offering "revenue-sharing." Founded in 2009, Demotix claimed to have in excess of 25,000 users, of whom more than 4,500 are active. It put a nice gloss on its interest in "beating the cost-cutting forces destroying international news. We want your journalism to make a difference. . . . Demotix was founded on the cross-roads of activism and journalism, with two principles in mind: freedom of speech and freedom of information." Similarly, Third Report spent a considerable amount of time explaining how contributors would "share in the fruits of their labor." Essentially, Third Men providing exclusive content receive half the advertising revenue their articles generate. But, like Demotix, it never said how much people have actually earned. Nonpaying sites were not any better at raising ethical consciousness. LocalByUs, founded and "self-funded" by one engineer—a Seattle-area mother of two—likewise never mentioned ethics.

NowPublic, to take one prominent example of a site concerned with relationships with other users and community guidelines, discussed its concern for the integrity of the site and community, which it protected by requiring users to abide by a code of conduct. Deliberate violations, it noted, result in loss of privileges. NowPublic's code forbade members from posting content in which they have a financial interest or with an explicitly commercial purpose. It also forbade plagiarizing; threatening other members; posting libelous, obscene, or pornographic material or content that contains racist, sexist, homophobic, and other slurs; and deliberately posting stories that are false or misleading. But, like MyNews, NowPublic requires members to be accountable for stories they post, correct known errors, and promote interactivity and dialogue that supports real community collaboration.

Several professional sites had affirmative definitions of good reporting. CNN iReport, for example, said: "It's yours (your own words and images); true, new, and interesting." But these definitions were thin and, again, accepted wholly unproblematically. In addition, CNN iReport did not explain its promise "to expand the current definition of news." Notably, iReport said stories need the basics (who, what, where, when, why, and how), and they need to be true and fair. It also said that a good story "connects" and "feels real." Whether emotion, described as "a powerful connector," might create panic or hysteria was never considered.

Discussion

Although we never expected that a single site would satisfy all criteria, we had expected to find many examples of a variety of best practices. Instead, for several dimensions, we found only one or two; for others, we found no examples. As a result, the best practices listed in Appendix 2 largely represent potentiality, not actuality. Few citizen journalism sites took ethics seriously, regardless of whether they are organized by mainstream news organizations, individuals with professional journalism, citizen groups, or individual citizens.

Ethics education, to the extent that it exists on these sites, falls into a few major categories. The news-sharing or news aggregator sites rely on users to moderate each other and build reputation through peer evaluation of comments or posts. Many traditional news organizations, as well as content-sharing sites, offer only a legalistic Terms of Use or User Agreement prohibiting activities. These banned activities are the kinds of "obvious" behaviors that are widely condemned and, in any case, are generally limited to behaviors that would result in legal actions against a user or the hosting site. For example, users of NPR.org's social networking tools are told to respect people's privacy and avoid personal attacks; hate speech and threatening, bullying, pornographic, sexist, and racist comments are prohibited. *The Washington Post,* for example, said that submitted content may not violate privacy or intellectual property rights; be libelous, defamatory, predatory, or sexually explicit; advocate illegality or violence; degrade others on the basis of protected categories such as race; or attempt to intimidate or harass. But what if people genuinely disagree on what is sexually explicit?

Community-run sites occasionally offer more elaborate guidelines and provide for checking by editors before posting. Nonetheless, again the emphasis is on prohibited behaviors—which are almost entirely behaviors broad enough that they should be applied to anyone online—rather than affirmative obligations or responsibilities specific to citizen journalists. Likewise, the legalistic definitions of prohibited behaviors are not elaborated or explained.

One surprise was that sites attached to universities or university-based news organizations were no better than the others in offering ethics education, and they were arguably worse. They simply ignore the opportunity for a teaching "moment" for inspiring or provoking consideration of ethics. Online since 2006, ChicagoTalks is a community and citizen journalism news website supported by Columbia College (Chicago). It explicitly endorses participatory culture and the Creative Commons. Ethics goes unmentioned. Likewise is My Missourian, the citizen journalism "grassroots" publication at the University of Missouri, with a journalism faculty well-known for commitment to ethics and to citizen journalism. According to the site, "We make every attempt to publish all original, local content within the bounds of civility." It goes on to define civility terms of four rules: prohibiting profanity, nudity, personal attacks, and attacks on race, religion, national origin, gender, or sexual orientation. This, and a promise that an editor reviews all content "with a gentle hand," completed what served for ethics. The Local East Village, operated by students and faculty at New York University's Journalism School in cooperation with *The New York Times,* fell short in much the same way other sites associated with universities did: It offered only a cursory explanation of why users ought to contribute and no discussion at all of what principles ought to govern the behavior of a user in contributing to the site. For its terms of use, the site defaulted to *The New York Times* policy, which is, like so many others, legalistic and negative in nature.

Given how easy it is to list some materials for further reading, even if a site prefers not to discuss them at length, a second surprise was that only three provided references to ethical resources. NowPublic did mention that it had consulted ethical resources and named a few it found inspiring. The Twin Cities Daily Planet provided an extensive list of resources for citizen journalists, beginning with the Poynter Institute and including the Center for Citizen Media. Mynews linked to several articles explaining the importance of citizen journalists in coverage of important news but not readings that addressed journalism ethics.

Professional journalists debate many ethical dilemmas. Why should citizen journalists not also address these questions? Moreover, since citizens are rarely graduates of journalism schools, why not offer access to some ethics education or at least provide links to resources? Even if posters do not necessarily regard themselves as citizen journalists (and they do not), these sites are explicitly providing platforms for citizen journalism and opening up the potential for citizen audiences to rely on them rather than more conventional news outlets. The standards on these sites define, at best, a bare minimum of appropriate behavior for users online. They suggest that it is fun and easy—at most a simple question of netiquette and avoiding what would seem to be patently wrong or even illegal behaviors.

Recent instances suggest that stronger ethics guidance is appropriate for citizen journalism. For example, the inaccurate report of Steve Jobs's death in 2008 by a citizen contributor to CNN's iReport led to a major drop in the stock market and an SEC investigation. These concerns are not unique to citizen journalism: *Bloomberg* mistakenly published Jobs's prewritten obituary in 2008, and CBS inaccurately tweeted his death a few months before he died in 2011. Still, professional journalists face these issues with the oversight of a news organization with an ethics code; they can discuss them in person with experienced colleagues. Citizen journalism sites, if they solicit contributions from citizens, should not only emphasize the importance of ethics but teach it. After all, as they continue to proliferate and generate more contributions, they will, and should, be taken seriously.

Why do citizen journalism sites fail to provide guidance or education in ethics? Would it cost them anything to do so? Many market-driven sites seem intentionally unreflexive and/or opaque about their strategy: The language suggests that they are seeking financial gain by appealing, perhaps falsely, to citizens who think they, too, can make a profit. We suspect that the sites run by mainstream professional news organizations, despite recent remonstrance to the contrary, retain a patronizing view of citizens—as unable to undertake serious work or serious thinking and thus perhaps beyond the reach of ethics teaching. Sincere-minded and good-hearted citizens who run community sites may simply never have thought about this and merely consulted attorneys for the terms of use language. Hybrid sites may have forgotten. This question itself deserves further study. In any case, in our view, citizen journalism deserves and requires much more robust and philosophically complex discussions of ethics—for the benefit of the sites and the communities they serve, as well as the citizens engaged in journalistic activities.

Appendix 1: The Sample

News/Content sharing (often communities):

- Slashdot (http://slashdot.org/)
- Reddit (http://www.reddit.com/)
- Digg (http://digg.com/)
- StumbleUpon (http://www.stumbleupon.com/)

- Topix (http://www.topix.com/)
- Twitter (http://www.twitter.com)

Totally citizen-run (citizen contributions without editorial oversight from pros):

- Allvoices (http://www.allvoices.com/)
- MyTown Colorado (http://www.mytowncolorado.com/)
- Digital Journal (http://www.digitaljournal.com/)
- Localbyus (http://www.localbyus.com/)

Citizen-run, with a user-moderated editing process:

- Kur05hin (http://www.kur05hin.org/)
- Wikinews (http://en.wikinews.org/wiki/Main_Page)
- Merinews (http://www.merinews.com/)

Citizen news for possible pay:

- Demotix (www.demotix.com)
- The Third Report (http://www.thirdreport.com/)
- Patch (http://www.patch.com)

Hybrid (editorial oversight, selection of content, often hyper-local):

- Salida Citizen (http://salidacitizen.com/)
- SierraBear (http://sierrabear.com/home/)
- Bluffton Today (http://www.blufftontoday.com/)
- WestportNow (http://westportnow.com/)
- Northwest Voice (http://www.bakersfieldvoice.com/)
- iBrattleboro (http://www.ibrattleboro.com/)
- MyNews (mynews.in)
- Now public (http://www.nowpublic.com/)
- Twin Cities Daily Planet (http://www.tcdailyplanet.net/)

University-connected local:

- Chicagotalks.org (http://www.chicagotalks.org/
- My Missourian (http://mymissourian.com/)
- East Village (http://eastvillage.thelocal.nytimes.com)—*New York Times* project

Corporate-owned, citizen contributions:

- MSNBC Newsvine (http://www.newsvine.com/)
- CNN iReport (http://ireport.cnn.com/)
- CBS EyeMobile (http://www.cbseyemobile.com/)
- Yahoo! YouWitnessNews (http://www.flickr.com/groups/youwitness-news)
- Washington Post (http://www.washingtonpost.com/)
- National Public Radio (http://www.npr.org/)

Appendix 2: Best Practices

- Include an accessible, readable code of ethics that states the importance of ethics affirmatively.
- Offer norms for what reporters do as well as what they do not do.
- Provide a clear, honest, but also ambitious self-definition.
- Be transparent and honest about site operations, revenue structure, ownership, number of active contributors and readers.
- Enable accountability of contributors to readers/site/community and of site/operators to readers/community.
- Provide a model of and for ethics, including by offering ethical education, and in terms of best practices.
- Address ethical dilemmas involving photos, graphics, video, audio, and sound bites.
- Require citizen journalists to read relevant material before registering or posting, separate from the terms of use or service.
- Complicate the notion of fairness, accuracy, and neutrality.
- Include ethics in self-regulation and in ratings of credibility and trust.
- Present citizen journalism as difficult, requiring much thought and consideration.
- Address relationships with sources and subjects and how to disclose this to readers.
- Define fairness not only as covering multiple sides of a story but also in terms of which stories are covered.
- Explain systems for monitoring and punishing violations of ethical standards.

- Consider underrepresented sources and citizens (especially children and elderly, as well as race, ethnicity, sexual orientation).
- Offer a broad and precise definition of diversity.
- Provide resources and citations for further thinking about ethics.
- Explain the importance of ethical citizen journalism, including with respect to democracy.

Notes

1. Apparently worried that it lacked specific purpose and editing consistency, in 2010 OhmyNews abandoned the practice of citizen journalism and launched a blog about citizen journalism.
2. See http://www.ojr.org/ojr/wiki/ethics/
3. See http://www.kcnn.org/principles/
4. Relevant here are the Poynter Institute, the Association for Practical and Professional Ethics, and the ethics committee of the Society of Professional Journalists (SPJ) as well as its Ethics Advice Line.

References

Allan, S. (2006). *Online news: Journalism and the internet.* Berkshire, UK: Open University Press.

Brennen, B., & Wilkins, L. (2004). Conflicted interests, contested terrain: Journalism ethics then & now. *Journalism Studies, 5*(3), 297–309.

Bruns, A. (2008). *Blogs, Wikipedia, Second Life, and beyond.* New York, NY: Peter Lang.

Castells, M. (2007). Communication, power and counter-power in the network society. *International Journal of Communication, 1*(1), 238–266.

Costanza-Chock, S. (2008). The immigrant rights movement on the net: Between "Web 2.0" and comunicación popular. *American Quarterly, 60*(3), 851–864.

Cyberjournalist.net. (2003). A blogger's code of ethics. Retrieved from http://www.cyberjournalist.net/news/000215.php

Daugherty, T., Eastin, M. S., & Bright, L. (2008). Exploring consumer motivations for creating user-generated content. *Journal of Interactive Advertising, 8*(2), 16–25.

Dowd, M. (2008, December 1). A penny for my thoughts. *The New York Times.* Retrieved from http://www.iht.com/articles/2008/12/01/opinion/edowd.php

Dowd, M. (2009, April 22). To tweet or not to tweet. *The New York Times.* Retrieved from http://www.nytimes.com/2009/04/22/opinion/22dowd.html

Friend, C. (2007a). Gathering and sharing information. In C. Friend & J. B. Singer (Eds.), *Online journalism ethics: Traditions and transitions* (pp. 54–79). Armonk, NY: M.E. Sharpe.

Friend, C. (2007b). Commercial issues and content linking. In C. Friend & J. B. Singer (Eds.), *Online journalism ethics: Traditions and transitions* (pp. 180–197). Armonk, NY: M.E. Sharpe.

Gant, S. (2007). *We're all journalists now: The transformation of the press and reshaping of the law in the Internet age.* New York, NY: Free Press.

Gillmor, D. (2004). *We the media: Grassroots journalism by the people, for the people.* Sebastopol, CA: O'Reilly Media.

The [Hutchins] Commission on the Freedom of the Press. (1947). *A free and responsible press.* Retrieved from http://www.archive.org/details/freeandresponsib029216mbp

Iggers, J. (1998). *Good news, bad news: Journalism ethics and the public interest.* Boulder, CO: Westview Press.

Johnson, B. K. (2008, May). *Incentives to contribute in online collaboration: Wikipedia as collective action.* Paper presented at the 58th annual conference of the International Communication Association, Montreal, Quebec.

Johnson, D. (1997). Ethics online: Shaping social behavior online takes more than new laws and modified edicts. *Communications of the ACM, 40*(1), 60–65.

Knight Commission on the Information Needs of Communities in a Democracy. (2009). *Informing communities: Sustaining democracy in the digital age.* Washington, DC: The Aspen Institute.

Kuhn, M. (2007). Interactivity and prioritizing the human: A code of blogging ethics. *Journal of Mass Media Ethics, 22*(1), 18–36.

Lemann, N. (2006, August 7). Amateur hour: Journalism without journalists. *The New Yorker.* Retrieved from http://www.newyorker.com/archive/2006/08/07/060807fa_fact1

Mill, J. S. (1869). *On liberty.* London, UK: Longman, Roberts & Green. Retrieved from http://www.bartleby.com/130/

Nov, O. (2007). What motivates Wikipedians? *Communications of the Association of Computing Machinery, 50*(11), 60–64.

Reich, Z. (2008). How citizens create news stories: The "news access" problem reversed. *Journalism Studies, 9*(5), 739–758.

Robinson, S. (2011). "Journalism as process": The organizational implications of participatory online news. *Journalism and Communication Monographs, 13*(3), 137–210.

Rosen, J. (2006, June 27). The people formerly known as the audience. *PressThink.* Retrieved from http://archive.pressthink.org/2006/06/27/ppl_frmr.html

Salter, L. (2009). Indymedia and the law: Issues for citizen journalism. In S. Allan & E. Thorsen (Eds.), *Citizen journalism: Global perspectives* (pp. 175–186). New York, NY: Peter Lang.

Schaffer, J. (2007). Citizen media: Fad or the future of news? *Knight Citizen News Network.* http://www.kcnn.org/research/citizen_media_report

Shirky, C. (2008). *Here comes everybody: The power of organizing without organizations.* New York, NY: Penguin.

Skube, M. (2007, August 19). Blogs: All the noise that fits. *Los Angeles Times.* Retrieved from http://www.latimes.com/news/opinion/la-op-skube19aug19,0,3547019.story

Surowiecki, J. (2004). *The wisdom of crowds.* New York, NY: Random House.

van Dijck, J. (2009). Users like you? Theorizing agency in user-generated content. *Media Culture Society, 31*(1), 41–58.

Woyke, E. (2009, March 11). The struggles of OhmyNews. *Forbes.* Retrieved from http://www.forbes.com/forbes/2009/0330/050-oh-my-revenues.html

Yang, H., & Lai, C. (2010). Motivations of Wikipedia content contributors. *Computers in Human Behavior, 26,* 1377–1383.

CHAPTER SIX

To Post or Not To Post

Philosophical and Ethical Considerations for Mug Shot Websites

MARK GRABOWSKI & SOKTHAN YENG

Newspaper websites are tapping into a curious cultural fascination: mug shots. You do not have to commit a serious offense, you do not have to be convicted, and you do not even have to be a public figure. Just get arrested—no charge is too small—and the head shot taken at the police station when you get booked could be indefinitely immortalized on your local newspaper's website (McLaughlin, 2008). These online photo galleries are growing in popularity as newspapers seek more web traffic and the advertising revenue that accompanies it. Small and major publications alike have joined the trend. While the sites are based entirely on information already available from local police departments, they do not tell the whole story, including who was ultimately convicted or who had charges dropped.

Journalists argue that the sites are just a continuation of their long-standing practice of reporting on local crime and those involved in it, a custom that dates back to the first colonial newspapers. Some publishers have even started lucrative newspapers and released books that feature nothing but mug shots (McLaughlin, 2008). Publishing such content online, however, raises new ethical issues that are unique to digital media, particularly for respectable newspapers, which are expected to "adhere to the highest standards of professional journalism" (Daily Press, 2011, para. 1).

There was a time when embarrassing information eventually died away. Or, if it did not, people could move and reinvent themselves. In the past, if someone got arrested for a minor offense, chances are you would not read about it in the newspaper or see the person's face on the evening news. At worst, there might be a brief mention of what happened in the police blotter—a record that would disappear after a week or so when the recycle truck came to collect old newspapers. But the Web has changed that. Now, individuals risk being branded negatively forever. As *The New York Times* observed in a story about disgraced former Congressman Anthony Weiner and negative online data, "The Web is like an elephant—it never forgets, and if let loose it can cause a lot of trouble" (Sullivan, 2011, para. 1). More and more prospective employers are using Internet searches to find information about people (Preston, 2011), and mug shot galleries often show up first on searches.

What is to be done about this? At the moment, anything goes when it comes to online mug shot galleries. There are no codes of conduct or best practices for sites to emulate. Even among newspapers, standards differ widely. Steve Myers, managing editor of Poynter, a renowned media studies institute, outlined some of these pressing issues that newspapers operating mug shot sites must grapple with:

> Is this journalism? Voyeurism? Entertainment? ... Is it fair to highlight people who have been arrested but not been convicted of a crime? What if the charges are dropped or they're acquitted? What are the legal implications of highlighting these people? ... In an age when things seem to live forever online, what impact could this have on people's digital identities? (2009, para. 3)

In addition to the media, online mug shots also raise issues for society, which can be illuminated through an analysis of the history of criminality, as outlined by French philosopher and social theorist Michel Foucault. Among philosophical issues we consider are these: How do these websites operate as yet another mechanism of social control and regulation? How do they contribute to the discourse about what constitutes acceptable and criminal behavior? What measures are available for resisting and/or reshaping an ideology that works toward heightened social surveillance? Answering these questions can help journalists design sites that better serve the public.

This chapter argues that these sites are not inherently unethical because they provide a valuable public service, which could be considered watchdog journalism. Newspapers could address much of the warranted criticism by adhering to some basic guidelines using the Society of Professional Journalists' code of ethics as a framework. Yet practical and philosophical issues remain because of the nature of these Internet galleries. In order to address the unseemliness of online mug shots, journalists and philosophers alike must consider the intimate connection between information and capital. How much does profit seeking drive news coverage? Is the modern subject increasingly seen as a generator of capital? Mug shot galleries highlight the tension around the connection of information and capital precisely because the accumulation of capital is often used as justification for the existence of such information sites. While the "good" and "bad" subjects have long been defined in relation to capitalistic concerns and ideals, mug shot galleries show that public education on this matter seems, perhaps more than ever before, tied to the flow of capital.

Growth of Mug Shot Web Sites

Although mug shot galleries are a hot new trend in digital media, they are not completely novel and they are not solely related to the Internet. Since 1950, the Federal Bureau of Investigation has maintained its "Ten Most Wanted List," a poster of mug shots prominently displayed nationwide in public places such as post offices (History Channel, n.d.). In 1988, Fox aired "America's Most Wanted," a weekly television broadcast featuring photos of fugitives and reenactments of their alleged crimes (Prial, 1988). It ran for 23 years, becoming the network's longest-running show (TV.com, n.d.). And for decades, *The Baltimore Sun* has been publishing a page in its print edition of police arrest reports that give names, addresses, and charges for people arrested the previous day (Myers, 2009, archived chat).

In the late 1990s, the mug shot phenomenon hit the Internet, when The Smoking Gun began curating a gallery of mug shots, featuring celebrities, infamous fugitives, and the "world's dumbest criminals" (Rector, 2008). The popularity of the content spurred entrepreneurs to create entire websites devoted to mug shots. Newspaper executives saw the money that could be made from such sites and joined in the fray. In the past few years, several major daily newspapers such as *Newsday*, the *Tampa Bay Times*, and *Chicago Tribune* have launched websites based on mug shots of recently arrested local residents. Even small newspapers, such as the 30,000-circulation Panama City, Florida, *News Herald*, now have online mug shot galleries. These sites have been wildly successful. *The Palm Beach Post*, for example, estimates its mug shots draw half of the newspaper website's 45 million monthly page views (Padgett, 2009, para. 3).

There is no archetype for mug shot sites. *The Palm Beach Post*, for example, shows the 100 most recent arrests in the region. Beyond the photos, there is scant information about the suspects' names, charges, and book time and date. Those featured on a recent day faced charges ranging from public drunkenness and driving with an expired license to grand larceny and attempted murder. No explanations of the charges or details about the circumstances surrounding the arrests are provided. *Newsday*, by comparison, provides a few paragraphs of details regarding the incident.

The *Chicago Tribune* is considered by many journalists to have the highest standards when it comes to mug shot websites (Smith, 2010, para. 3). The newspaper only runs mug shots if there is a staff-reported story to accompany it. "We have a mug shot and a caption and we link a story to that caption," explained Bill Adee, the newspaper's digital editor. "That really sets the bar. We use mug shots where there are stories to set the context" (Smith, 2010, para. 8). But content is not limited to high-profile miscreants such as convicted Governor Rod Blagojevich or murder suspects. The *Tribune's* mug shot website relies heavily on TribLocal, the newspaper's chain of suburban websites, which describes itself as a "unique mix of professional and user-generated content (Miner, 2010, para. 12)." Out in the suburbs, murder is not necessary to get some ink. On a typical day, visitors to the mug shot site may see alleged child beaters, panderers, and burglars. "We try to get different kinds of crimes and we try to get a variety of city and suburban," Adee said (Smith, 2010, para. 3). *Chicago Reader,* an alternative weekly newspaper, observed the following about the *Tribune's* mug shot site: "'Mugs in the News' samples the local criminal element, choosing mugs with the same careful regard for the overall effect as a florist assembling a bouquet" (Miner, 2010, para. 13).

Some newspapers have much more controversial practices. The *Tampa Bay Times,* for example, occasionally includes mug shots of juveniles on its website. "The youngest I've seen was 16," recalled staffer Matt Waite, who designed the site. "He was accused of shooting someone and had been charged as an adult. As such, we do not filter them out. That was the subject of intense discussion" (Myers, 2009, archived chat). Many sites, such as the Newport News, Virginia, *Daily Press,* allow visitors to make comments on each mug shot—a feature one public defender calls "online Salem pillories" (Padgett, 2009, para. 5).

The information on the sites is not always reliable. Sometimes, police make mistakes or provide misinformation. "Several of the [police departments] warned us that the data they input can been flawed," Waite said (Myers, 2009, archived chat). In other cases, people featured on online galleries have been innocent or later acquitted of charges (Smith, 2011, para. 3). Most websites do not follow up to see how the case played out in court (Myers, 2009, archived chat). Many innocent people may not be aware they are on these sites, unless someone tells them or they actively look for themselves.

Some sites will freely take down photos upon request if the person was wrongly arrested, the charges are dropped, or the case leads to an acquittal. But several websites charge a fee to remove content, regardless of whether the person was acquitted or convicted of the charges. In some cases, fees can be as high as $400 (Kravets, 2011, para. 6). Some websites—including those operated by newspapers such as the Ogden, Utah, *Standard-Examiner*—by policy do not remove mug shots from their sites if the defendant has been found not guilty or their record has been expunged (Horiuchi, 2011, p. 3). "We post only true and factual information as originally published by local law enforcement agencies," an official at MugShots.com said in defense of the policy. "We make no judgment, we take no sides" (Horiuchi, 2011, p. 3).

Policies also differ widely for the period of time the photos are displayed online—a major source of controversy. Some websites store the content indefinitely, which means innocent people may have their mug shots google-able for the rest of their lives. Other sites, such as the *Tampa Bay Times* (Myers, 2009, archived chat) and the *Chicago Tribune,* remove the content after 2 months. "So they're not in that gallery forever," Adee said. "And if charges are dropped, we drop the mug shot" (Miner, 2010, para. 14).

Some newspapers also take measures to prevent their website content from being indexed by search engine bots, reducing the chances that a link to the mug shot will appear when someone does a search for the arrested person. Waite explained as follows:

> I can remember a conversation that we had very early on where I said that I did not want the first record in Google to be our site for anyone. So we've taken steps to stop Google from indexing the individual pages . . . we've worked very hard to make sure that this information is not Google-able. (Myers, 2009, archived chat).

But many sites do not take such measures, and some even utilize search engine optimization practices to ensure their content appears first if someone does an Internet search for the name of a person in their mug shot gallery (Kravets, 2011, paras. 2, 5).

Despite all of these questionable practices, it is "legally safe" to publish the mug shots, meaning website operators cannot be sued for publishing what is essentially a government document, according to lawyer John Watson, an associate professor of journalism at American University (McLaughlin, 2008, para. 13). George Rahdert, a public records lawyer who represents the *Tampa Bay Times,* said even if the person in the picture is found not guilty, he or she was still arrested and, under the law, the newspaper has no legal obligation to remove the mug shot from its website (Smith, 2011, para. 7).

Additionally, most mug shot websites use legal disclaimers to cover themselves. A boilerplate warning is prominently featured on many websites, including the *Chicago Tribune's,* stating the following: "Arrest and booking photos are provided by law enforcement officials. Arrest does not imply guilt, and criminal charges are merely accusations. A defendant is presumed innocent unless proven guilty and convicted" (Smith, 2010, para. 17).

Website operators are not the sole antagonists in this controversy. These online galleries would likely not be possible if not for police cooperation. When The Smoking Gun began its mug shot section, it gathered its information only after filing tedious, time-consuming Freedom of Information Act requests (Rector, 2008, para. 8). Now, many police departments not only share arrest records with mug shot websites, but in some cases they even seem to encourage them to use the information (Ashford, 2010). Computers in the *Tampa Bay Times* newsroom, for example, automatically download arrest records and mug shots each day from the servers of Tampa Bay area police departments (Myers, 2009, archived chat). A number of police departments also operate their own mug shot galleries on their websites. Consequently, some critics say media are just a symptom of the problem and that law enforcement is to blame for the problems caused by online galleries. Not all police departments release mug shots. The New York Police Department, for example, releases photos and arrest records only if they are actively searching for a person (Ashford, 2010, para. 4).

Mug Shots as Part of Foucault's History of Criminality

The emergence of mug shot websites can be seen as an extension of the history of criminality as developed by Foucault. The French philosopher traced a trend, starting in the 18th century, which moves away from gruesome corporal punishment to discipline through compulsory visibility of the criminal. Foucault argued that harsh physical injury to the offender is unnecessary; instead, a system of signs needs to be put in place that will allow the criminal to be more readily identified and make the general public more aware of its own possible transgressions. Imposing discipline on the public is more effective if it is pursued through the minds of the population than on the physical body of any particular criminal (Foucault, 1995, p. 101). What must be put on display are not the markings of corporal punishment but the idea of what constitutes the "good" citizen and the "bad" criminal. This system of signs and representations, as Foucault showed, cannot be separated from the growth of knowledge and capital in modern society.

Thus, the development and growth of websites devoted to the display of mug shots should not be a surprise. This content, especially on the Internet, is an offshoot of the information revolution. Since knowledge is increasingly measured by the amount of information available at one's fingertips, the education of the public via representations becomes more and more the norm. In fact, no physical harm or actual crime need occur. Mug shot galleries are sustained by the image of the criminal and a cultivated public interest in deviant behavior. However, these picture galleries do not thrive simply because they offer entertainment for the morbidly curious or undisclosed information about a neighbor or potential employee. They must also be understood within the framework of social knowledge of the criminal. Defenders of mug shot galleries can claim that they not only provide a public service but also serve to circulate knowledge about the state of criminality in modern society. People who create, search, or browse mug shot galleries are informed about what does or does not qualify as acceptable behavior. Websites which display mug shots, therefore, follow in the path of ideological education of deviant behavior by providing and proliferating the image of the criminal. Viewers, ideally, can also take part in shaping the discourse about criminal acts and the definition of the criminal. For instance, employers do not have to reject potential hires solely on the grounds that their pictures were featured on a mug shot website. Those who find information on these websites can also make a judgment about which (alleged) offenses sufficiently criminalize the accused. For example, a neighbor or employer may respond differently to someone who is featured on a mug shot gallery for driving with an expired license than to someone who is charged with driving under the influence. So while these websites do not necessarily predetermine the image we have of our community members or coworkers, these sites are—for better or worse—part of a system that seeks to publicize the lines between socially acceptable behavior and those behaviors that are transgressive and dangerous.

Publicizing Crime and Punishment

The system and archive of knowledge of the criminal created through mug shot galleries cannot be denied. Even calls for and measures to make mug shot galleries less controversial and/or more palatable to the general public can be seen as part of the project to educate and shape how the public should view criminal behavior. While following up on Internet mug shots by providing information about convictions or acquittals may save editors from some criticism, it does not solve the philosophical problem. Further reporting on the accused (who is convicted or acquitted or what sentences are doled out) may make these sites less controversial. These efforts also bring them in line with the philosophy that seeks to effectively administer ideas about criminal behavior.

Citing legal theorists, Foucault (1975/1995, pp. 95–96) highlighted the need to draw lines clearly between crimes and punishments. Coupling criminal activity with the punishment that is meted out helps the public to understand what is and is not accepted by society.

> This general element of certainty that must give the system of punishment its effectiveness involves a number of precise measures. The laws that define the crime and lay down the penalties must be perfectly clear, "so that each member of society may distinguish criminal actions from virtuous actions." (as cited in Brissot, 1781, p. 24)

> These laws must be published, so that everyone has access to them; what is needed is not oral traditions and customs, but a written legislation which can be "the stable monument of the social pact," printed texts available to all; "Only printing can make

> the public as a whole and not just a few persons depositories of the sacred code of the laws." (as cited in De Beccaria, 1764, p. 26)

The Internet, in general, and, particularly, mug shot galleries that include follow-up on punishment received, have the potential to go further than any other previous medium in disseminating information that will allow the greater public to distinguish between criminal and virtuous activity. Mug shot galleries—unlike, say, the web pages of *The New York Times*—do not place a limit on the viewable information that the general population (those without a subscription) can access. Because these websites invite perusal, the masses largely have unfettered access to most mug shot galleries. The growth in popularity of such sites also means the greater public is aware of legal codes that delineate criminal behavior. Additional information about those featured on the website serves to further enforce the legal code by clarifying the links between certain criminal behavior and the given punishment. Follow-ups on a featured mug shot will make it clearer for the public to see what punishment will be incurred for each crime.

The Economy of the Criminal

Beyond combatting crime, these sites can provide additional utilitarian value. In order for a general circulation newspaper to succeed, it must appeal to a variety of reader interests. The mug shot galleries are not simply useful because they deliver what the public desires—embarrassing details about their neighbors' personal lives—but also because they generate revenue that can fund news stories. Delivering the latter depends on success of the prior. Thus, increased traffic to mug shot websites would be useful for all who work in the field and for the public at large. This highlights how closely information and profit are connected in the digital world, but the definition of the criminal has long been connected to moneyed interests.

Foucault suggested, " . . . the economy of illegalities was restructured with the development of capitalist society" (Foucault, 1975/1995, p. 87). The rise of capitalism and the garnering of wealth helped shape the idea of criminal behavior and script responses to criminal behavior. Because capitalism ushered in greater wealth, the rise in property crimes increased. Accordingly, the economy of crime shifted so that a higher moral value was placed on property than previously existed (Foucault, 1975/1995, p. 87). With capitalism came greater intolerance against economic crimes and increased measures to ensure the security of property and wealth. Thus the rise of capitalism brought with it a new order of criminality and a new understanding of the human subject. Increasingly, the human subject became defined in relation to capital.

While there is a case to be made that mug shot galleries perform a needed function, we should nevertheless be aware of the possible dangers that lurk when the subject is increasingly defined and seen through the economic prism. This goes both for those who are featured on the website and the journalists who provide the stories. Undoubtedly, abolishing these sites would lead to the unemployment of many journalists and others in the media industry. And the dwindling of journalists also means that there will be fewer resources to do reporting. But is this argument just another extension of the capitalistic logic? Capital trumps both the journalist and those featured on the mug shot galleries. We should keep in mind that the more information and capital are linked, the greater the number of people that will be subsumed by the circulation of capital.

The Internet showcases how the human being came to be laden with economic value. The idea that humans provide economic value and are producers of surplus labor value is not new. However, the display of alleged criminals on Internet mug shot galleries carries out the logic of alienated labor in such a way that we must, as professor of Media Studies, David Golumbia, suggested, not only question the dream of capitalism but of the acquisition of information.

Golumbia applied a Marxist critique of capital to the way in which the Web can further exploit humans for the profit that they generate. While Marx made clear in *Capital* (Marx & Engels, 1906, p. 249) the ways in which those who work in factories of one kind or another produce profit for the capitalist, Golumbia applied this analysis to the exploitation of the human subject on the Web (Golumbia, 1995). Marx sought to show the value of the laborer that is hidden through capitalism. Golumbia, in turn, argued that the World Wide Web has the power to further obfuscate the figure of the human laborer who creates capital. We suggest that mug shot galleries are exemplary cases of web-based profiteering and offer some solutions to capitalistic exploitation, for the Internet allows for a dangerous blending of human subjectivity and information for the purposes of increased consumerism and ever-increasing capital.

Golumbia suggested that the circulation of capital as seemingly divorced from the human subject on the Internet is unprecedented. "The World Wide Web offers a startling new instance of this process of circulation, and especially of the ways in which capital itself uses the process of circulation to create forms that exist 'independent of the individuals'" (Golumbia, 1996, p. 26). The exploitation of the human subject or creator of capital can become ever greater because subjects may not choose or even be aware of the ways in which their thoughts or images—in the case of mug shot galleries—are used to generate profit and circulate capital. The subject becomes transformed into web-based artifacts that can be searched in the attempts to obtain information and, in turn, circulate capital.

On the Internet, humans can be flattened and made to serve the dictates of the website or cyberspace that represents their ideas and/or personhood. Thoughts and images of human beings may seem to be completely divorced from the capital they generate, especially if the subjects are not willing participants of or contributors to the webpage. The tendency to read the thoughts of others and see the images of people as packages and products to be consumed is multiplied through the Internet. Golumbia called on us to recognize how the Internet functions within the circulation of capital as "... a powerful construct built upon the lives and blood of real persons (ourselves included), whose labor becomes the stuff of capital through direct exploitation and through the processes of alienation" (Golumbia, 1996, p. 36). Once the packet of information and commodified meaning is thought to exist independently of the individual, the circulation of capital will reach an ever-greater level of exploitation of human subjects as the distance grows between themselves and the profit they produce.

Applying SPJ's Ethics Code to Mug Shot Websites

Mug shot websites unquestionably offer benefits to both newspapers and society. But there are also significant problems surrounding the sites. This section addresses whether major criticisms and philosophical dilemmas can be remedied using traditional ethical practices utilized by journalists. Specifically, we apply the Society of Professional Journalists' (SPJ) code of ethics, an ethos voluntarily embraced by thousands of writers, editors, and news professionals. It has four major principles: Seek truth and report it, minimize harm, act independently, and be accountable (Society of Professional Journalists, 1996).

Journalism or Voyeurism?

Critics argue mug shot galleries are more voyeurism than journalism. "The public at large loves to see people degraded," said Salt Lake City criminal defense attorney Ron Yengich, who called the sites "despicable":

> We have become a very mean society ... without mercy ... that doesn't understand the presumption of innocence at all. It does not add anything to the public debate about crime and how we deal with crime. It just gives the citizenry at large a way to make fun of people. (Horiuchi, 2011, p. 4)

But proponents point out that entertaining is part of the journalism business. Indeed, many journalism textbooks teach that to "entertain, inform and educate" is the ethos of the newsroom (Harcup, 2007, p. 5). Criticizing newspapers for running mug shots galleries to increase website traffic is akin to criticizing a reporter for crafting a story people will want to read, proponents argue. Ryan Chief, who operates *Busted*, a magazine and soon-to-launch website featuring mug shots, said the following:

> It's America's Most Wanted and COPS and probably a little Jerry Springer in there too. I'm not claiming I'm Clark Kent or Bruce Wayne. There's obviously a sensationalized factor as well . . . for the same reason people watch COPS . . . or for that matter read the *L.A. Times* or the *Chicago Tribune* for an article about someone arrested for a DUI or a sexual assault. (Ashford, 2010, para. 22)

There should be limits, however, to entertaining. Newspapers are institutions that are valued for their credibility and, as such, must maintain certain standards of decency. As the SPJ's code of ethics implores, "Show good taste. Avoid pandering to lurid curiosity" (1996). Certain features on the sites, such as allowing visitors to leave anonymous comments, overstep the bounds of good taste by inviting ridicule. The distinction between essential information and the gratuitous is just one standard for determining what shows good taste.

SPJ's code does not simply call on journalists to be tasteful about which particular information is included for each mug shot, but it can also be a call for journalists to think about the way such sites are organized in general and how they contribute to class and cultural stereotypes. Questions about the arrangement of mug shots (like a floral bouquet, by displaying those deemed ugliest, dumbest, etc.) show how journalists are embroiled in a larger discourse about taste. These sites help to define what is "vulgar" and "tasteful" within society. It is not a coincidence that sites which conspicuously violate the SPJ's code regarding taste trade on existing stereotypes. Thus, special sections and specialized websites, such as "Texas's Ugliest Criminals," "Dumbest Criminals of the Week," and "32 Mug Shots That Will Make You Crap Your Pants"—a sampling of the types of sites one may find doing a Google search for "mug shots"—arguably contravene SPJ's guideline. While the industry would certainly have even lower revenues without mug shot galleries, borders of taste should also be recognized. Newspapers should not sell their souls to stay in business; otherwise, they risk becoming tawdry tabloids instead of respected sources of information.

Fairness and Balance

Another major criticism of mug shot galleries is that they are unfair because they do not tell the entire story. First, most sites provide only scant details about the arrest. This is problematic because the first tenet of SPJ's code of ethics is to "Seek the truth and report it" (1996). Watson, of American University, argued that mug shots could be just as likely to mislead a reader as they are to inform if they are missing important details about the arrest—where, when, and how the alleged crime occurred and whether it was serious (McLaughlin, 2008, paras. 25–27). For example, he cited "public lewdness," a charge that could mean a range of things, from urinating in public to exposing oneself to children. A photo captioned simply "public lewdness" might wrongly influence a reader to think the person pictured is a pervert when the individual merely could not find a public restroom and relieved himself or herself in a back alley when a cop happened by.

Another unfair practice, critics say, is that most sites do not follow up on the charges—was the person convicted? McBride, of Poynter, said because the mug shots only tell an incomplete story, they violate journalistic ethics.

> It's unethical to report an arrest and never follow up. What's the journalistic purpose of doing that? Mostly it's for prurient purposes. . . . And it can cause great harm to the individual who is wrongly arrested for something horrible like assault, but never actually charged. Imagine if that's the first thing that comes up on your Google search. (Myers, 2009, archived chat)

While journalists should always strive for accuracy in their reporting, some mistakes are inevitable. Newspapers would never be published if perfection were required. This is why newspapers are often said to be the "first rough draft of history." Even before the Internet, there was a long-standing tradition of reporting on crime using only police as the source. Oftentimes, initial arrest reports were never further investigated. Ken Walker, reporter at the *Tampa Bay Times,* said the following:

> It's common practice for newspapers to run crime blotters—lists of calls to which law enforcement officers responded, whether or not an arrest was made. Is that unethical as well? There's no follow-up, and a call indicates even less about guilt than an arrest. Yet, it's useful information about what's going on in a neighborhood. Is that unethical, too? (Myers, 2009, archived chat)

For a limited period at least, mug shots and the charges against those depicted in them represent our best approximation of the truth. Accused criminals do not instantly go to trial; in some jurisdictions, it can take up to several months following an arrest for a case to commence. Provided that websites remove mug shots before the trial begins, the information can be said to be timely and valid—based on the police reports. But steps should be taken to prevent the content on such websites from being indexed by search engines. Otherwise, the mug shots could continue to be accessible in search engines through cache memory long after they have been removed from the website's server.

McBride conceded that requiring newspapers to follow up on all initial arrest reports represents a departure from the accepted norms in journalism.

> I realize that the industry practice has been different. And indeed when I was a police reporter I didn't follow up on everything. I can't defend that. Because if it's important enough to spend the resources putting into the paper, you should be dedicated enough to make sure your audience knows the whole truth about the matter. (Myers, 2009, archived chat)

McBride is not alone. Many critics of mug shot sites insist newspapers should report on how the charges played out. "I think it would be far more useful to pull court records and report on convictions and sentences," crime victim advocate Tina Trent said. "You could run the mug shots once somebody has been actually found guilty" (Myers, 2009, archived chat).

While such a proposal may seem fair, it also raises issues. Tracking a criminal case as it moves through the court system can be tricky for even experienced lawyers. Fewer than one in 40 felony cases now make it to trial (Oppel, 2011, para. 5). When deals get struck and charges get dropped, news releases do not often get issued (Miner, 2010, para. 15). Even when a judge or jury issues a verdict, the outcome often does not mesh with what actually happened. For example, a conviction may be the result of inadequate legal representation, a faulty witness, or planted evidence. An acquittal does not necessarily equate with innocence. It just means that the prosecution could not convince a jury beyond a reasonable doubt.

A possible compromise to this quandary of how to tell the "whole truth" is to give those charged a chance to clear their names in the media. In other words, there should also be a means to allow the accused to respond to—Foucault might say resist—the categorization of miscreant. Mug shot sites should have a mechanism to remove people who were acquitted or had charges dropped. SPJ's code states, "Be accountable.... Admit mistakes and correct them promptly" (1996). Removal fees for the innocent and acquitted should be eliminated. In addition, sites should give innocent and acquitted people the option of being featured in a special section proclaiming the dismissal of the charges against them. Edward Wasserman, journalism professor at Washington & Lee University, explained as follows:

> An acquittal or a dropped case is essentially no different from a correction, and if media organizations, to their credit, are more aggressive now in setting right often trivial errors they make, they ought to bring the same zeal toward clearing innocent people of baseless reputational harm that they, in the normal course of doing what they consider their duty, have done a great deal to cause. (Miner, 2010, comments)

Even those who are convicted should have an opportunity to tell their side of the story. This is consistent with SPJ's code, which states the following: "Diligently seek out subjects of news stories to give them the opportunity to respond to allegations of wrongdoing" (1996).

Right to Know Versus Personal Privacy

Finally, site operators must balance the public's right to know with personal privacy. In the case of mug shot galleries, audiences are seeking information about private individuals that those individuals would rather not have disclosed. So, site operators must weigh the relative importance of two ethical principles—providing information that will help the public make decisions and respecting an individual's right to privacy. Certainly, it is embarrassing to have your mug shot online for friends, family, neighbors, and coworkers to see. But mug shot galleries also clearly offer a form of watchdog journalism.

Becky Bowers of the *Tampa Bay Times* argued as follows:

> It does help the community learn about itself. People have mentioned they had no idea so many people are arrested each day in our area—and on such an array of minor issues. Certainly I was surprised to see an 86-year-old woman arrested on suspicion of writing bad checks. I wondered if the traditional criminal justice system was the most appropriate venue. In other words: There's a transparency to this that's taught me more about the machinery of justice in my community than I'd noticed from reporting of the "big deal" crimes. That's valuable to me. (Myers, 2009, archived chat)

Chief, of *Busted,* agreed. "This is public information. . . . It informs communities of recent arrests," he said. "By doing so, we're educating them on what's going on in their community" (Ashford, 2010, para. 8). Chief cited one instance in which a mug shot he published helped identify a sex offender at a YMCA in a Michigan community. It prompted background checks that identified 22 registered sex offenders who were not supposed to have been allowed membership (Ashford, 2010, para. 8).

Waite added the following:

> The main journalistic purpose of this feature is that we've given transparency to the grinding wheels of the justice system. The jail population is no longer an abstraction. You can look at them, as they come in. These people are your neighbors. The jail, the deputies that run it, the courts that have to deal with these folks, you pay for it. So there is a purpose to showing that to people. I would also add that people have said they found great value in being able to look at people who said they lived in a specific ZIP code because they only know their neighbors by sight. (Myers, 2009, archived chat)

Overall, mug shot publishers seem to be on sound ethical footing. The SPJ ethics code states that journalists have an obligation only to the public. "Act independently. Journalists should be free of obligation to any interest other than the public's right to know" (1996). But there is a difference between what the public wants to know and what it needs to know. While there is value in informing the public about a child molester or drug dealer in the neighborhood, it is questionable if there is significant value in featuring people charged with minor misdemeanors such as driving with an expired license. Not all crimes are equally newsworthy. SPJ's code advises journalists to "minimize harm." But this does not mean to avoid harm completely. Some individuals will be harmed in the form of public shame and ridicule so that the community may be safer. But, minor offenses, perhaps, should be overlooked. In addition, juveniles should be left off the sites. As the code notes, "Use special sensitivity when dealing with children and inexperienced sources or subjects."

Conclusion

Like them or not, mug shot websites will likely continue booming as newspaper executives look for new sources of revenue. In fact, the day may come when everyone knows at least one person who has appeared on such a site. Although controversial, there is no denying that mug shot websites offer some benefits to society—even if their owners are operating them for purely opportunistic reasons. Mug shot galleries help define who is a "normal" and "good" member of society. As Foucault would concede, not all normalizing powers are bad. He stated, "My point is not that everything is bad, but that everything is dangerous, which is not exactly the same as bad" (Foucault, 1983, p. 231). If posting a mug shot of an alleged drunk driver helps to enforce the idea that drunk driving is bad, this would help create a positive norm or standard of behavior.

However, there are some questionable practices in the industry that urgently need to be addressed or newspaper websites risk being likened to tabloid journalism. The danger is that news editors contribute to class and cultural stereotypes through mug shot galleries. There are, however, mechanisms that allow criminal identities to be challenged or, at least, not be overdetermined by these websites. If the information is not provided to large search engines or if the information can be disputed or changed, the accused have a greater ability to mold their public identity and work against the injustices of negative typecasts.

By implementing a handful of measures, mug shot websites may be able to alleviate many ethical concerns. Specifically, limiting the duration for storing mug shots on servers and preventing content from being mined and indexed by search engines will help prevent those featured in online galleries from having their digital identities permanently tainted. It will also ensure that newspapers of record are not perpetually publicizing information that may be incorrect or outdated, a betrayal of their mission. Creating a mechanism for innocent people to clear their names and allowing everyone featured on the site to offer an explanation will provide fairness and balance, two staples of journalism ethics. Banning comments on mug shots may help those pictured avoid gratuitous ridicule from the community. Explaining charges could lessen the likelihood of readers making false assumptions and spreading misinformation. Finally, newspapers should consider whether all types of criminals, regardless of seriousness of offense and age of offender, should be included in their database. While there are merits to being egalitarian in covering crime, not all crimes may be newsworthy.

Of all these suggestions, limiting the searchability of mug shot galleries may trouble news editors who champion transparency. These limitations, however, do not signal that the news sector is ignoring its responsibility to provide information to the public. To the contrary, news editors who place limits around the searchability of their mug shot galleries could help to create a culture of more careful and responsible information gathering. Performing a quick general search for a particular person's name will not immediately uncover his or her mug shot, if it exists. Yet those who seek specific information about another's criminal record would still be able to find it on the newspaper's website. Because the information can be found at a site connected to a newspaper, there will be a greater chance that further information about the alleged crime (or its dismissal) can be provided. Operating mug shot galleries in this manner also has the benefit of redirecting web traffic to the newspaper's website. If mug shots cannot be found on larger search engines such as Google, those seeking such information must go to the newspaper source. Greater web traffic allows newspapers to attract more advertisements to their site and build funds for further news reporting.

References

Ashford, M. (2010, February 26). Mugshot publishing continues to expand. *iMediaEthics.* Retrieved from http://www.imediaethics.org/index.php?option=com_news&task=detail&id=752

Brissot, J. (1781) *Theorie des lois criminelles* (Vol. 1). Paris, France.

Daily Press. (2011). Statement of journalistic ethics for the Daily Press, Inc. News Department. Retrieved from http://www.dailypress.com/services/dp-ethics,0,2832688.htmlstory

De Beccaria, C. (1819). *Of crimes and punishments* (E.D. Ingraham, Trans.). Philadelphia, PA: Philip H. Nicklin (original work published in 1764).

Foucault, M. (1983). On the genealogy of ethics: An overview of work in progress. In H. Dreyfus, P. Rabinow (Eds.), *Michel Foucault: Beyond structuralism and hermeneutics* (2nd ed., pp. 228–255). Chicago, IL: The University of Chicago Press.

Foucault. M. (1995). *Discipline and punish: The birth of the prison* (A. Sheridan, Trans.). New York, NY: Vintage Books. (Original work published 1975)

Golumbia, D. (1996). Hypercapital. *Postmodern Culture, 7,* 36.

Harcup, T. (2007). *The ethical journalist.* London, UK: Sage.

History Channel. (n.d.). This day in history: Mar 14, 1950: The FBI debuts 10 most wanted. Retrieved from http://www.history.com/this-day-in-history/the-fbi-debuts-10-most-wanted

Horiuchi, V. (2011, September 6). Your mugshot is online; it could be gone—for a price. *Salt Lake Tribune.* Retrieved from http://www.sltrib.com/sltrib/home2/52450098–183/com-mugshot-mugshots-florida.html.csp

Kravets, D. (2011, August 2). Mug-shot industry will dig up your past, charge you to bury it again. *Wired.* Retrieved from http://www.wired.com/threatlevel/2011/08/mugshots/ ethics/ Content?oid=2888360

Marx, K., & Engels, F. (1906). *Capital: A critique of political economy.* New York, NY: Random House.

McLaughlin, E. (2008, December 8). Media taking mug shots—foreign, familiar—to bank. *CNN.* Retrieved from http://www.cnn.com/2008/CRIME/12/08/mugshots.fascination/index.html

Miner, M. (2010, December 16). Doing face time: The ethics of a burgeoning trend in journalism. *Chicago Reader.* Retrieved from http://www.chicagoreader.com/chicago/mugs-in-the-news-journalism

Myers, S. (2009, April 8). Archived chat: The ethics of posting mug shots online. *Poynter.* Retrieved from http://www.poynter.org/latest-news/top-stories/95107/archived-chat-the-ethics-of-posting-mug-shots-online/

Oppel, R. (2011, September 25). Sentencing shift gives new leverage to prosecutors. *The New York Times.* Retrieved from https://www.nytimes.com/2011/09/26/us/tough-sentences-help-prosecutors-push-for-plea-bargains.html?pagewanted=all

Padgett, T. (2009, September 21). Newspapers catch mug-shot mania. *Time.* Retrieved from http://www.time.com/time/magazine/article/0,9171,1921604,00.html

Preston, J. (2011, July 20). Social media history becomes a job hurdle. *The New York Times.* Retrieved from http://www.nytimes.com/2011/07/21/technology/social-media-history-becomes-a-new-job-hurdle.html

Prial, F. (1988, September 25). Freeze! You're on TV. *The New York Times.* Retrieved from http://www.nytimes.com/1988/09/25/magazine/freeze-you-re-on-tv.html

Rector, K. (June/July 2008). High-caliber ammunition: The Smoking Gun makes its mark online with its relentless pursuit of documents. *American Journalism Review.* Retrieved from http://ajr.org/Article.asp?id=4545

Smith, D. (2011, September 23). Should mug shots really be posted online? *My FOX Tampa Bay.* Retrieved from http://www.myfoxtampabay.com/story/18027827/should-mug-shots-really-be-posted-online

Smith, S. (2010, December 17). Is *Chicago Trib*'s mug shot publishing more ethical than *St. Petersburg Times*? *iMediaEthics.* Retrieved from http://www.imediaethics.org/index.php?option=com_news&task=detail&id=1658&search=mug%20shot

Society of Professional Journalists. (1996). SPJ code of ethics. Retrieved from http://www.spj.org/ethicscode.asp

Sullivan, P. (2011, June 10). Negative online data can be challenged, at a price. *The New York Times.* Retrieved from http://www.nytimes.com/2011/06/11/your-money/11wealth.html

TV.com. (n.d.). America's Most Wanted. Retrieved from https://www.tv.com/shows/americas-most-wanted

CHAPTER SEVEN

Shaping Our Shadow

Erin Reilly

We each have the ability to build a networked cultural identity through the communities we choose to participate in and the profiles we shape or allow others to shape in the social networks we choose to belong to. However, what about children who grow up within networked families who have a collective construction of identity? What happens when children transition to adolescence and want to represent themselves differently than how their families have shaped their identity?

According to an AVG Digital Diaries: Digital Birth 2010 survey, the average age children acquire an online presence is 6 months, with more than 70% of mothers posting baby pictures online and sharing them through social networking sites. (Brand, J. and Ronner, R, 2010) Where some new parents have learned to make videos private, sharing with only a select few, others have not learned this practice or have chosen to go public from the start. Just search the term "sonogram" on YouTube and thousands of videos fill the page, posted by men and women eager to claim their new identity as parents-to-be. Click on one of the sonogram videos, such as hearing the heart beat for the first time, and you'll notice that it has been viewed by thousands of people. Comments range from personal, where the mother is reflecting 9 months later on the video, to random people who can identify with that moment in time as a form of nostalgia. It's obvious from the comments that not all of the thousands who have viewed this one video are friends and family.

Though we cannot identify who the child is, he or she has become part of a public mediated performance of how we shape childhood today.

Posting a sonogram for all to see is the start of the screen becoming the child's "shadow," a representation of identity, similar to how the shadow of Peter Pan is represented in J. M. Barrie's story (Barrie, 1981). At the beginning of *Peter Pan,* Peter's shadow somehow falls off and becomes free, detached from its original source . . . but not for long once Wendy captures it and sews it back on to Peter. We can't escape who we are or where we are along our journey, as shown in the constant theme in *Peter Pan* of the trials and tribulations of being a child and coming of age into adulthood. Peter's freed shadow jumping from bed to bed, changing shape, color, and mass, is a tangible representation of Peter's identity.

The screen becomes a child's shadow, as real experiences become performances shared and shaped by the public. As we free our "shadow" throughout networked communities, many rather than one create control of the process by which we shape our identities online.

Online social networks, forums, and blogs provide rich opportunities to network, communicate, and share information with vast audiences. At the same time, disclosing information online can be harmful to those involved if they fail to consider the ways in which the information they share about themselves and others could be used. danah boyd described four properties of online environments, which suggest the need for caution. These properties include the persistence, searchability, and replicability of information posted online and the presence of invisible audiences (boyd, 2007). Given these properties, the everyday decisions we make regarding what to disclose, to whom, and how, become urgently important, especially when it comes to shaping the identities of our children. In the early stages of development, children do not have control of their identities that are being created and represented by those they love. However, shadows that are hidden over time can persist and come out of the dark when light is shed upon them, and therefore, we cannot see children as passive participants when constructing who they are before they have a voice in the process.

Ethical Considerations for Parenting Publicly

When thinking about identity formation, it's important to explore our own conceptions of privacy and those of others. For all of us, there are both promises and risks associated with these opportunities. By creating a profile and sharing at least some personal information online, one can participate in small, private, online communities or large, public networks.

A recognition of these varying perspectives is important for youth and adults in thinking about what level of privacy is appropriate for them, the consequences of their decisions of posting, and the considerations they must make in order to respect the privacy wishes of others.

Take, for example, the infamous YouTube meme, David after Dentist. YouTube user booba1234 posted a video of his son, David, who is high on laughing gas after having had his tooth pulled at the dentist, where he then, in the backseat of the car, poses philosophical questions such as, "Is this real life?" and "Will this be forever?" According to booba1234, David was extremely nervous about going to the dentist so the father decided to shoot some video of before and after in order to help calm David's nerves. The father also has shared that his wife wanted to go but had to work, so as is the case in today's tele-cocooning family, videotaping the experience to share later has become the next best thing for families to "shadow" the experience (http://www.davidafterdentist.com). Within a week of posting, David's YouTube video had over five million views. It had gone viral with many remixes of the original. Different talk shows invited David and his father to come on and retell again and again the story which allowed for commenting and showing it to a larger audience than those who saw it just on YouTube.

Although the content of the video is seemingly innocuous and David's family has received monetary rewards that will help support David's future, such as his college education, the question of whether it was fair to David for his father to share this intimate moment with the world needs to be asked, and people did just that through comments on the YouTube video that pointed to David's father being inattentive and the Bill O'Reilly show agreeing with these outlying comments. No matter if the choice is to post or not to post something about a child to the public; people will form pictures of who we are on the basis of the information they find. Information is interpreted and reconfigured through the eyes of the beholder, and these interpretations are layered onto the representation of our identity that we share with the public.

This content seems to be an asset to David's family both socially and monetarily. An everyday life experience from an ordinary family has turned their child into a "star" of the Internet world where people want to take photographs with him and have David sign autographs. David is a new type of child star emerging through playlists on YouTube. But what if David ends up not liking the media attention, as he gets older? He might rephrase his original question that put him into the limelight to, "Is this real or just a shadow of who I am?" A majority of his child identity is made up of one incident that has been taken up, remixed, and shaped by the masses. When does the child have a right to control the flow of information about his or her identity, especially when it is already out there and cannot be deleted?

As children grow into adolescence, the family identity transforms toward explicit interaction where teens begin to understand the abstract concepts of their own identity and the makeup of the family's group identity. They are more aware and concerned with how they and their family are being acknowledged by others and thus want a voice in how that is shared and interpreted. Without a deeper understanding of the different stages of child development, are families aware of boundaries that might be overstepped that could have repercussions later in life? What at the time was funny could extend into being an embarrassment for adolescents, especially if children's view on privacy and how they want to shape their online identity is different than that of their parents.

Family Identity Goes Public

When our homes become a public space for others to peek into, what happens to privacy and how is our collective identity as a family shaped and represented by others? The screen provides us a voyeuristic opportunity to peer into other people's lives, perhaps for comparison to our own identity or just because we are curious. Parents need to be in the habit of reflecting on the potential consequences, for themselves and for their children, of such decisions, because once it is shared via the screen, we can never fully extricate ourselves from history.

This holds true for parents who include their children in the family's public identity. If children are at an early stage of development, they may not be aware of what it means to have a private moment and are entrusting themselves completely to their caretaker. In a sense, parents view their children as extensions of themselves; therefore, sharing images and video of their children can be interpreted as a performance of their own identities as mothers or fathers.

ShayCarl from *The Shaytards* is a good example of the extension of his own performance onto his children. His byline on his website states, "My kids think I'm funny! And they're cuter than your kids." (http://shaycarl.com/) This statement holds true with an accounting of him, his wife, and his kids through daily videos you can subscribe to, giving you the opportunity to shadow this family of five as they video document their life and share it with the public via their vlog, which is a form of blogging through video that has been around since 2005 and is often considered "web television."

You'll notice on *The Shaytards* YouTube channel that tens of thousands of people subscribe to them for a lens into the life of this family. In 2009, ShayCarl was nominated as one of the top 10 vlogs on YouTube, thousands of people tune in each day to find out more about this family. One fan shared this: "Shaycarl and family is the best! Best family EVAR! :) even the kids are funny and outgoing plus shay and mommytard really gives a good parenting to their childs. shay ftw!" (http://www.the-top-tens.com/lists/best-youtube-vlogs.asp) The multitude of videos on the YouTube channel are about daily life, whether it's the Shaytards going fishing to shopping for PrincessTards sixth birthday to hanging out at home, learning how to do a back flip.

A vlog such as the Shaytards has put the power of creating the story into the hands of the family, or more specifically, the parents who are making the choices as to what to shoot and what to keep private and held within the walls of their home. However, many ethical concerns arise through the Shaytards offering an unscripted, often unedited account of their lives and with little rules yet established on the type of content allowed via the multitude of video and photo server sites.

One of the first questions people ask is, "Why does ShayCarl, the father, name all his family members with the suffix of–tard?" Many people assume that it is a shortened reference to the derogatory epithet "retard" and find the choice offensive. In actuality, however, the–tard suffix is given for other reasons, including to safeguard the family by having each person go by a nickname to avoid revealing their real names but if you are new to the series and do not have this frame of reference, the naming can easily be misunderstood and taken out of context. The other issue is when you are so prevalent throughout the Web, though the thought was good, in reality, fans have searched the video archive and due to unedited material, each person's real name has been revealed—which brings into question the ethical dimensions of an outsider's participation in forming one's identity.

More heated controversy about the parenting style of the Shaytards abounds on social networks. One video of ShayCarl states that he became a vlogger so that he wouldn't have to get a real job, but the content that he receives money for often revolves around the exploitation of his children, such as having his 1-year-old "maintain" a Twitter stream—@rocktardstweets.

Then there is the question of what is the tipping point when we move from roughhouse playtime to child abuse, and are we taking it out of context because we have more or less information provided? ShayCarl is a boisterous, big guy and loves to throw and wrestle his children . . . and on occasion, with any wrestling match, a bump in the nose or a bigger fall than expected can occur. The majority of people would be just as concerned if they saw this same incident in a public park, but it exponentially grows in controversy due to the publicness of the accusations and the ability to circulate and have a wider net cast for more to wag their fingers. With the Shaytards shooting their every action shared and replayed and put together in a collection, this can be misrepresented and grow into cause and concern in the home.

The choices made today by parents Katielette and ShayCarl shape their children's public identities now and for the future. It is their choice to participate in a *Truman Show*-esque existence, in which their family identity is linked to the public's opinion of what it is like to be in this family. Within a family, social power is implicit during the early stages of childhood, with the parents maintaining control and responsibility. Often, the first place a child's identity is formed is at home with the family; parents are the first teachers, the influencers of their children. However, with family being mediated by technology, we must question whether parents understand the power that they wield and what type of representation of the 21st-century family is being projected and accepted as normative.

Online Identity Play: Possibilities and Problems

So how do we grant youth greater control over their public identities? Through digital media, youth use photos, interests, "favorites" lists, and other content to play up—or hide—different aspects of their identities. They engage in what many scholars refer to as "identity play"—in which they explore and receive feedback on new identities (for example, a more confident self) or develop facets of the self (sexual or gender identities, for example) that they may not feel comfortable exploring offline. Mediating these practices through flexible digital technologies allows for ample opportunities to reinvent oneself as desired or needed.

All of us are involved in identity play to some degree, though in the real world, we may have fewer options to radically transform who we are or to sit back and review the performances we enact. We bring forward different aspects of ourselves as we interact with the various people in our lives or as we move through different contexts. Increasingly young people are deploying new technologies as resources to help them manage these different roles and facilitate different aspects of their identity.

A vital part of growing up is developing one's identity. For children like David or Rocktard, they both have had their digital identities formed by their parents in public. The promise of these scenarios is that identity play will come more naturally to them because their identity so far has always been in relation to the online world. They also will have parents who will be more understanding of their actions as they make mistakes and learn from them. However, at the same time, this could become a disadvantage, making it harder for either of them to manage and extricate the shadow that will follow them as they move into adolescence. As teens, they also might not see the point of using digital media as a space to try on new identities, which many adolescents view as part of a coming of age in today's networked culture. One possibility for this potential lack of interest is that they might have a stronger awareness of how identity online can easily be tied to exploitation and monetization from their experience with their family.

On the other hand, youth who have not had as long of a shadow have more opportunities to form their own biographical sketch. Take, for example, Sam, a 13-year-old from Illinois, who was chosen by Edutopia to participate in a contest. Edutopia is George Lucas's nonprofit organization, whose mission is to support innovation and reform learning. The organization was requesting teens to submit videos to their *Digital Generation* project that represented who they were. This gave Sam the opportunity to personally design and submit a video that represented her own digital biography. Rather than having someone else create her identity online, Sam had the digital know-how to creatively showcase her interests. Unlike David and the ShayCarl's kids, whose identities were first crafted by their parents, Sam was given the opportunity to consciously construct alternative identities for herself and fluidly move between her online and off-line lives to emphasize different characteristics of who she is through use of game avatars, video edits, sound and image manipulation. Through this submission, Sam wrote herself into being (Sundén, 2003, p. 3). Writing oneself into being is a way to project an identity out into the digital world so that others know whom you are and can interact with you via the multitude of media you portray yourself through. This can be as simple as having a profile on Facebook to having multiple social networks that you use for varied identities, as well as YouTube, Flickr, and Twitter spaces to connect with others. There are some die-hard digerati who do not think you exist if you have not written yourself into being.

Sam carefully constructed her being through the design of her video and profile through a playful portrayal that introduces her as Sam, I Am, a creative 13-year-old who not only eats a banana and introduces her BFF but also highlights her Wii remote, reports on some science news with Ditto and Jiffy, and in a transformed voice, introduces us to her character on *World of Warcraft.* She had control over the aspects of the identity she put forward. She can pick and choose what aspects of herself to highlight and which to downplay, enacting what Erving Goffman has called "impression management" (1959) and influencing how others perceive her.

Sam enacted the right identity since she was chosen as one of the teens to represent Edutopia's *Digital Generation* project. But this is where "impression management" is removed from Sam as a single author of her identity to one that is written by many. In being chosen to represent the digital generation, Edutopia producers invited Sam as well as others close to her, including her parents and teachers, to expound upon Sam's biographical sketch, portraying her identity in a more formal documentary, interview-style persona. They then synthesize, edit, and re-envision a new Sam to share with the world in relation to the other teens who were also chosen.

Now and always, Sam will be known as part of this collective identity for others to reference in better understanding of what today's teens are like growing up with digital media. Her individual sense of self is marked in relation to this group where many others like her claim that digital media is their "second life," using this metaphor to represent how her entire digital identity is similar to how her virtual character in Second Life can move freely through the virtual world, interacting with others and participating in a culture that she relates to as much as hanging out with her friends in real life.

Digital Footprints and Implications for Identity

With ubiquitous access to others and easy access to participating in varied communities, the question of how we communicate ourselves to the world becomes more important than ever before. The lines between our public and private lives have blurred with the rise of Facebook, Twitter, Foursquare, and other social networking sites. Whether it is our immediate family or a community we identify with and belong to, it is not only ourselves who make choices in how we sculpt our identity but also those around us—so consideration of others becomes more important in this connected world, especially to those who do not have a voice or are too young to understand the implications.

Some forms of online identity exploration can be deceptive, undermining relationships and causing distrust in communities, or even closer to home, in families themselves. Even within a home, harm can be done if these identities are performed, even if the intention is not to deceive. Harm can happen when the multiple identities we develop clash into each other.

Take, for example, timid, socially awkward Jessica, who wanted to try on new identities and reinvented herself online as sexy gothic model, Autumn Edows. Jessica was one of many children portrayed in PBS's Frontline television show, "Growing Up Online: Just How Radically Is the Internet Transforming the Experience of Childhood?" Where Jessica walked the school hallways alone, Autumn gathered fans following her every move. Through this experience, Jessica (aka Autumn) grew more confident, yet she kept this identity secret, and it was by others' hands that her parents ended up learning about her fame, resulting in her two worlds crashing into each other.

As Jessica discovered through the creation of Autumn, information about identity posted online can often be seen and interpreted by people who we would never encounter face-to-face. Reality is skewed by how we shape the media to portray our identities, thus resulting in others not having the full picture of who we are. Consider, for example, the assumptions about identity that one can make based on information posted to a social networking site such as a Facebook profile. The photos, notes, and updates accessible through the site give a particular, fragmented picture of identity that is open to interpretation. Combined with fewer opportunities to explain the meanings of our varied identities to anonymous or semi-anonymous viewers, control over ways in which others understand our identities is drastically lessened. As a result, the image we portray might mean different things to different people.

Blending of fact and fiction is common in creating media and gives people the ability to shape their own identities. You've probably heard the saying, "Can you ever really know someone?" We must have a bit of skepticism in our viewing because it is likely that only partial representations are shared of a person rather than the whole picture.

This discursive process of constructing identities is not totally under our control but also shaped by people around us for their own reasons. Even when motives are benign, some negative consequences can occur. And this process raises ethical implications both in terms of whether we are truthful in constructing our own identities and whether we are ethical in granting youth greater control over the public identities we structure for them.

The choice of what to post and what not to post is really in the eye of the beholder, and when it comes to children, this belongs to the parent until a certain point in the child's development. Taking into account the child's personality and considering what the child would do if he or she were older and better understood the formulation of identity helps parents make the right choices.

Each identity we develop becomes a shadow that lurks in the background until light is shed upon it. Our voyage of discovery in the 21st century will have to be with new eyes, ones that are open and aware of the perception others may already have of themselves due to the multitude of identities created not only by oneself but also by others. One shadow quickly multiplies by the light of the fire, and others on the other side of the screen become dependent on these shadows to make sense of the world.

References

Barrie, J. M. (1981) *Peter Pan*. New York, NY: Random House, Inc.

boyd, d. (2007). Why youth (heart) social networking sites: The role of networked publics in teenage social life. In D. Buckingham (Ed.), *Youth, identity and digital media* (pp. 7–9). Cambridge, MA: MIT Press.

Brand, J. & Renner, Rona. (2010). *Digital birth—celebrating infants and toddlers online and offline*. In AVG Technologies Digital Diaries. Amsterdam, The Netherlands.

Goffman, E. (1959) *The presentation of self in everyday life*. New York, NY: Anchor Books.

Sundén, J. (2003). *Material virtualities*. New York, NY: Peter Lang.

Websites

AVG 2010 survey: http://shine.yahoo.com/channel/parenting/most-2-year-olds-already-have-a-digital-footprint-2397728.
ShayCarl Top Vlog: http://www.the-top-tens.com/lists/best-youtube-vlogs.asp
Shaytards name: http://www.shaytardsfan.com/2010/05/what-are-shaytards-real-names.html

Videos

David After Dentist: http://www.youtube.com/watch?v=txqiwrbYGrs
Jessica/Autumn: http://www.pbs.org/wgbh/pages/frontline/digitalnation/relationships/identity/the-secret-online-life-of-autumn-edows.html?play
Shaytards YouTube channel: http://www.youtube.com/user/SHAYTARDS?blend=1&ob=5
Sam: http://www.youtube.com/watch?v=FUnWE9UmKj0 and http://www.edutopia.org/digital-generation-profile-sam

SECTION 3

Subversive Uses of Online Spaces

INTRODUCTION BY ADRIENNE MASSANARI

Perhaps the most challenging issues facing our use of digital environments involve those on the margins—the so-called subversive uses that many of us engage in but few acknowledge. From texting sexual content to "griefing" in virtual worlds to sharing copyrighted content to engaging in online activism, new technologies invite a multitude of behaviors that are potentially unethical. The chapters in this section investigate these new uses of digital media and interrogate our assumptions of what constitutes "inappropriate" behavior in online environments. In particular, they suggest that we examine these actions without resorting to polemic or reductionist arguments about potential harms—especially as many of these actions are merely extensions of already-existing social practices.

Jo Ann Oravec's investigation of sexting, the exchange of sexual images and messages using mobile devices and computers, highlights the contradictory discourses that exist around many activities within digital media environments. In her chapter, "The Ethics of Sexting: Issues Involving Consent and the Production of Intimate Content," Oravec suggests that sexting is often framed as a moral issue, with clear-cut distinctions between victims and perpetrators. However, she notes that both policy and educational responses fail to adequately consider the fact that this practice is an extension of normal desires for contact between intimates. Oravec argues for a shift in our collective discourse around sexting, one that deemphasizes our moral outrage regarding the practice and instead offers practical advice for how to engage in texting ethically and safely.

Roland Wojak's "Griefing Through the Virtual World: The Moral Status of Griefing" examines the practice of "griefing" or actively and deliberately causing grief or annoyance to others within virtual spaces. Griefers often justify their behavior as meaningless or ultimately not harmful, as it takes place wholly within the game space. Wojak argues, however, that the distinction between "real" and the "virtual" spaces is increasingly blurry. Therefore, bullying and griefing behaviors online have real-world consequences, even if those consequences are not physically realized. At the same time, what behavior constitutes "griefing" remains underdeveloped, challenging both virtual world creators and inhabitants to craft appropriate community policies that protect individuals from this kind of bullying behavior.

Like Oravec, Brian Carey argues for a more nuanced understanding of our online activities that may be characterized as subversive or detrimental by the press and society at large. In his chapter, "Permissible Piracy?," Carey argues that there may be ways to share copyrighted content ethically (if not legally). He interrogates the generally accepted position that the sharing of copyrighted material is equivalent to stealing, instead suggesting a more nuanced view of piracy is needed. In particular, ethical frameworks do not fully account for times when a product is no longer available for purchase (as is the case with many out-of-print computer games, for example) or those moments when digital products may be copied for personal use.

Perhaps one of the most contentious and controversial organizations online is the subject of Alex Gekker's chapter, "Legionnaires of Chaos: "Anonymous" and Governmental Oversight of the Internet." Anonymous, an international, loosely connected group of hackers, embraces an ethos of Internet vigilantism, alternatively turning its collective focus on the Scientology church and law enforcement agencies hoping to shut down Wikileaks, a whistle-blowing organization that releases sensitive government and defense documents. Anonymous is emblematic of the techno-libertarian values that shaped the earliest days of the Internet's development, and yet its actions often raise the ire of law enforcement organizations and governmental bodies as it presents fundamental questions with regard to what information should remain secret (if any) in the age of global information sharing.

Digital media and new technologies will continue to present new and intriguing ways for us to engage as individuals and communities. However, not all of our uses of these technologies will be prosocial or ethical. We must understand these "subversive," and perhaps unsavory, activities because they say something about our darkest desires and how they may shape our relationships with others and our global community.

CHAPTER EIGHT

The Ethics of Sexting

Issues Involving Consent and the Production of Intimate Content

Jo Ann Oravec

Introduction

The Invention of Sexting

The joke "I'd like to meet the persons who invented sex and see what they're working on now" has had assorted iterations in topical conversation. "Sexting" may indeed be the answer to that question, an invention that fuses sexual interaction, technology, and communications. Sexting has been framed in cartoon-like caricature through incessant exposure in broadcast and print journalism (as well as late-night comedy) in the past several years; it has been a topic of considerable interest in online forums as well. The word "sexting" was entered into the 12th edition of the *Concise Oxford English Dictionary* (along with "cyberbullying") in August 2011 (Boyle, 2011). It has been considered seriously for DSM status (*Diagnostic and Statistical Manual of Mental Disorders,* which organizes diagnostic criteria for various declared mental disorders) as psychologists attempt to understand and interpret the behavior of adults employing cameras and computer capabilities in their sexual activity (Wiederhold, 2011). Extensive public attention to the development and exchange of sexual images and condensed narrative via mobile device (including smart phones and cell phones) has congealed a set of social practices into an object that attracts considerable social and political attention.

"Cybersex," the conveyance of sexually oriented material via computer networking, has been discussed in popular news conduits and social forums for decades. However, transmissions of suggestive images and text via smart phone or computer (once considered largely youthful stunts) have been transformed into a set of interactions widely associated with shame, negative publicity, and often illegality, whatever emotional content they contain for participants. Sexting has become a form of "toxic" communication, with sender, receiver, interveners, unintended recipients, social media platform managers, and even bystanders often being tainted in its aftermath. Such toxicity provides a useful function in some organizational contexts. For example, it equips news outlets that want to attack certain politicians and celebrities to do so, despite the fact that sexting practices are widespread and generally legal. Sexting has produced major ramifications for a number of its partakers, including resignations from high positions, divorces, felony convictions, and even suicides.

This chapter attempts to characterize the ethical dimensions of the monitoring of public discourse about sexting by those who participate in it, an activity that is tightly coupled with matters of establishing awareness and consent among parties engaged in these practices. Sexting itself is not an especially cerebral function and is often conducted in the heat of sexual passion. However, the context of sexting is information based and requires some level of observation on a broader scale for participants. Such monitoring can involve complex processes of gleaning information from sources including newspapers, blogs, legal opinions, and educational materials. Staying current as to what dangers sexting can place on individuals is somewhat akin to checking weather reports before going sailing on a large lake or bay. One can certainly face the prospects of sailing in rough waters with heroism and even excitement, but for best results, checking the weather reports is advised; obtaining the consent of one's crew before setting out is socially responsible conduct. The spectrum of activity placed under the rubric of sexting in the context of recent public discourse is expanding dramatically, as described in upcoming sections, and narratives are emerging from a variety of sources. In hundreds of small town newspapers across North America there are accounts of adults in various positions of authority who were caught sexting either on or off the job who were relieved of their duties or otherwise embarrassed because of a racy text or e-mail message. Eraker (2010) stated that the popularity of sexting "has outpaced both the law and technological limitations of prior generations" (p. 556), leaving in its wake a number of damaged reputations and diminished futures.

What are some of the ethical concerns involved with sexting? This chapter focuses on the activity of obtaining the information needed to establish meaningful and reasonably informed consent, given that the aftermath of sexting can include considerable negative consequences (which is indeed quite often). Consent issues have special dimensions when public awareness of sexting practices as well as social and legal responses to the practices are evolving at an accelerated pace. Sexting often involves social media platforms, including Twitter and Facebook, transforming intimate communications into those with potentially broader interactive spheres. Uncertainties concerning the platforms have added complications as to how these intimate social exchanges are conducted and construed for various societal purposes. For example, social media controls such as privacy and security settings have undergone a series of changes in the past few years, leading to confusion as well as disquiet (Penny, 2011). These issues take on special significance when toxic communications are involved in which the various parties handling or otherwise involved with the messages have the potential to be affected negatively in some way. Consent becomes more convoluted as well when there are considerable uncertainties about the technological context; despite the best attempts of individuals to keep material in social networks and electronic mail systems private or otherwise under control, leakages and manipulations of various sorts can and often do occur, so requests for consent could be rooted in empty promises of confidentiality. With a misplaced keystroke or the efforts of hackers, materials can be made public that were intended for private consumption. Just having digital photos on a smart phone or computer system can make them subject to release, as was alleged by actress Scarlett Johansson, who claimed in August 2011 that nude images of her were pirated from her phone (Risling & Jablon, 2011).

Sexting in an Age of Ubiquitous Photography

Sexting is one relatively small part of a panoply of forms of digital image generation. One of the legal characterizations of sexting in U.S. context is the following: "the practice of sending or posting sexually suggestive text messages and images, including nude or semi-nude photographs, via cellular telephones or over the Internet" (*Miller vs. Skumanick*, 605 F. Supp. 2d at 637). A number of individuals play direct parts or have potential roles in sexting, including senders, potential receivers, image or text modifiers, and retransmitters, as well as viewers who have acquired access to the material through surveillance. Other parties involved include those who work to characterize sexting as an immoral activity, creating a set of well-labeled societal victims and perpetrators and even fostering forms of moral panic. Many of these negatively characterized individuals are young people. Eraker (2010) asserted that, "in at least ten states, authorities have publicly contemplated, and in some instances followed through with, charging the teenage participants with child pornography production, possession, and distribution" (p. 556). Young people have had many personal images taken in a wide assortment of contexts (including full body scans in airports and surveillance cameras in dressing rooms in department stores), so the notion of being portrayed without clothing by advanced technological media is certainly not foreign to them. This is not a generation that associates the camera with "soul stealing" (Robinson & Picard, 2009) or other phenomena. Providing young people with cell phones equipped with cameras has often been problematic, giving them the opportunity to take photos of each other in various stages of undress in gymnasiums, sometimes without consent of some of the chosen subjects. Some school districts have countered these measures by banning the use of cameras in locker rooms ("Phone ban," 2011).

Monitoring public discourse on sexting, watching for changes in privacy protections on social media platforms, establishing channels for consent among partners, and otherwise attempting to mitigate harms associated with sexting would elucidate but still not eliminate the risks involved. Risk has indeed often been construed as part of the appeal of sexuality as well as one of its drawbacks. The notion that one can get easily caught in a dalliance and one's sexual interactions revealed through misplaced sext messages has been a factor linked to excitement as well as fear in many fictional and real-life sexual narratives. Many young people report sexting while driving, compounding whatever risks are involved in either activity (Harrison, 2011). The very notion of risk involves at least some awareness of context, although the modes through which many individuals receive information increasingly block out certain kinds of news and current event coverage. Assumptions that sexting participants know the current set of consequences for their activities and can judge risk in appropriate manners may be problematic, given the paucity of news and public affairs consumption observed among many individuals (Custódio, 2011).

Complex issues emerge even in simple cases of sexting, which require some level of societal monitoring to clarify and elucidate them. In the "standard" and relatively straightforward mode of sexting, an individual sends a message containing images or narratives that are sexually charged. The digital images produced are of the individual or individuals who is/are sending the messages, with little or no modification of the material. No one knows when or where the first individual "invented" sexting, that is, had the idea to send a racy photo to another individual via Short Message Service (SMS) or via the Internet, for that matter. Modification of digital images in itself has a long history, providing issues for journalists and marketers as well as amateurs for decades (Oravec, 1998). The kinds and varieties of sexting practices have expanded dramatically beyond the standard model; a substantial assortment of narratives about sexting has filled traditional news media as well as social media and has served to influence public policy related to sexting (for example, the wide assortment of bills related to sexting introduced in state legislatures in the United States). Consider the following description of the negative outcomes of sexting:

> Adolescents who send such pictures often realize too late that they have no control over where the images go. . . . They circulate among schoolmates, leading to ridicule and taunting. . . . Those were the consequences Jessica Logan of Cincinnati faced after sending nude photos of herself to her boyfriend. . . . The 18-year-old committed suicide two months after appearing on television in 2008 to teach others about the dangers of sexting. (Hicks, 2011, p. B01)

Participants in even the standard cases of sexting include the immediate producers and consumers of sexually related content, possible virtual "bystanders" or other audiences (such as those associated with the social media platform involved), as well as the surveillance personnel of social media platforms (including larger venues such as Facebook and more restricted settings such as blogs). The decision to send a message with personally identifiable sexual content to another individual may seem direct and unproblematic. However, complications may soon occur. For example, at age 16, Melanie Young sent a digital photo of herself without clothing to a male friend that was subsequently distributed widely online: "I thought it was fun and just a way of flirting. . . . I sent it to someone that I thought I could trust" (Goodson, 2011, p. 1A). Melanie Young was not alone in her choice of activity:

> One-fifth of the teenagers surveyed last year by *CosmoGirl.com* said they had transmitted or posted nude or semi-nude pictures of themselves online. Since the sample was drawn from teenagers who had volunteered to participate in surveys, it may have been biased toward exhibitionists. But the results suggest that high schools across America are rife with child pornographers. (Sullum, 2009, p. 16)

Receivers can be placed in considerable legal and reputational peril simply for receiving sext messages that incorporate sexually themed images of minors, as in the case related here:

> On a typical day Ben Hunt and his best friend John Eicher, both 14, send each other about a dozen routine text messages. But on Jan. 15, while at school, Hunt sent Eicher something on his cell phone that suddenly put their futures in peril: a photo of a girl in their class exposing a breast. Eicher didn't see it as a big deal. "I really didn't think of deleting it," says Eicher, an eighth grader who at the time attended the Lawrence School in Falmouth, Mass., with Hunt. "I was, like, 'Whatever.'" (Hewitt & Driscoll, 2009, p. 111.

Hunt and Eicher were subsequently involved in lengthy legal proceedings and faced dismissal from their schools. A particularly troubling case of sext message receivers who were placed in difficulty by the toxic aspects of sexting was reported by the U.S. National Educational Association (NEA). A teacher who attempted to intervene in a sexting case that potentially involved bullying was himself disciplined by his school district and by legal authorities for handling pictures of nude minors (Simpson, 2011). Receivers can include those who are employed by the social media and telecommunications purveyors that support the communications (who are sometimes directly associated with academic institutions). In some instances, individuals have been construed both as creators and receivers of sext messages:

> . . ."sexting" has been around, as a prank and a problem, for years: in 2004 a 15-year-old Pittsburgh, Pa., girl was charged with sexual abuse of children and dissemination of child pornography when she posted nude pictures of herself online. This seemed like a confounding twist in prosecutorial philosophy, since the victim and the villain in this case were the same child. But just in the past year, more than a dozen states have followed suit, arresting kids as young as 13 for sending or receiving smutty pictures on their phones. For parents, these cases have suddenly raised the prospect of retirement savings melted down to pay legal bills, college dreams deferred, scholarships lost—all because their kids were caught doing what kids do, and were prosecuted aggressively in hopes that others would notice and clean up their act. (Gibbs, 2009, p. 56)

Recipients are faced with the decision of whether to modify the text or image in some way; often, the sexual dimensions of the images are digitally modified in this stage. Manipulation of images so as to distort or enhance their sexual aspects has been made easy through such programs as Adobe Photoshop. Resending of sexts is yet another dimension of sexting. The exchange of sext messages for "insurance" purposes (in case of the dissolution of the romantic relationship involved) has been documented as well, as young people construct a sort of "Mutually Assured Destruction" (MAD) pact:

> Teenagers are exchanging explicit photos of themselves as a form of "insurance"' to stop their pictures being circulated via mobile phones. A two-year research project into sexting—sending sexually explicit messages, photos or videos via text message—shows adolescents are blackmailing each other with nude photos to protect themselves from their own increasingly risky behavior. (Lentini, 2011, p. 16)

The innovative resolution of intimate issues described earlier signals that individuals are making some creative efforts to solve problems involving common understandings and consent in the realm of sexting. Such "insurance" strategies are not permanent and stable solutions to sexting issues, but they at least provide for some level of empowerment for those involved in these practices. The insurance strategies also place genders in roughly equal roles, although very often the risks of exposure are still different in quality (given current differences in sexual standards related to gender). The strategies are akin in some ways to prenuptial agreements in which parties prearrange for their positions in case of dissolution of a marriage.

The process involved in developing such arrangements is often unsettling in that the agreements introduce ideas that might not have otherwise occurred to the couple involved.

Incorporating money into sexting interactions adds another layer of complexities and new parties to the discourse. Stokes (2010) described a number of bleak scenarios concerning sexting, including the use of digital camera feeds on the Internet for prostitution:

> . . . the "Cam whoring" phenomenon . . . presents the most difficult test case for researchers and the most difficult challenge in terms of ethical and legal research. The goal is to illuminate a potentially very harmful aspect of adolescence for "digitally native" people. One such person is Justin Berry. Berry is the most prolific "Cam Whore"—having made hundreds of thousands of dollars before he was 18 years old by performing on webcam, at first from his bedroom, later from an apartment rented for the sole purpose of performing, in exchange for gifts and money. (p. 319)

Although the kind of scenario just described is relatively rare, it still provides impetus for those who want to curtail and even punish sexting. It provides strong narrative themes that permeate social discourse on sexting and can influence public policy.

Other mutations of sexting practices involve adultery. Popular discourse on whether sexting constitutes adultery provides some clues as to the social construction of "sexting" (Wysocki & Childers, 2011). Such discourse has been particularly widespread in cases of celebrity and political sexting, in which the possible consequences of sexting for the spouse of the individual involved (if any) are often examined in detail in various media outlets.

Discussions concerning adultery are serving to expand the notion of sexting and place it in broader ethical and social context; often they inject titillating or amusing aspects as well that are captured in newspapers and blogs for the entertainment of readers. For example, some narratives in popular discourse place sexting as disruptive of marriages. In "Sexting Not Harmless Fun for the Cheated Wife," O'Riordan (2010) portrayed sexting as a form of spousal cheating:

> Which is exactly what secret social networker British TV presenter Vernon Kay told his wife Tess Daly. Kay was accused of "Twitter cheating" on his wife by conducting private online flirtatious conversations with a number of female acquaintances. In an attempt to explain his behaviour, Kay said the messaging started out quite "innocent" but quickly developed into something more explicit. He thought the messages were no more than "harmless banter"—that was until his wife found out about his overly social networking and helped change his mind.

As the variations and popularity of sexting expand, so does its integration into everyday married life. Godson (2010) commented on the question, "Is 'sexting' my husband a good idea?" in *The Times* of London. The individual posing the question stated that she had "no idea where to begin" doing so.

Celebrity sexting is yet another mutation. In the past few years an intense focus has been placed upon the phenomenon of sexting connected with a number of celebrity and political controversies both in the United States ("Anthony Weiner's Judgment," 2011) and in other Western nations (Juntunen & Valiverronen, 2010). Sext messages have been involved in a number of political, sports, and entertainment controversies. Several of the major recent sexting cases reported in news and entertainment venues, including that of U.S. football star Brett Favre, have popularized the procedures involved in sexting in detail. The public was informed of details of how and when the photographs were made and what images they incorporated. In some of the cases, including those of ex-Congressmen Weiner and Lee, media coverage often disseminated the digital photographs associated with the incidents to the public in print, broadcast, and Internet journalism. These sexting events have had consequences not just for celebrities themselves but also for fans and political supporters, a number of whom build psychosocial relations with major public figures (Fowler, 2011). Many celebrities increasingly endeavor to engage in interaction with their fans and supporters via social media such as Facebook and Twitter, often in ways that have very little to do with sex. However, the immediacy and mobility of sexting provides even more intense forms of fan engagement for some notable figures. The potential for sexting can be construed as placing the "fan" more readily into a sexual status in relation to the "celebrity." In turn, association with celebrities has often convoluted the status of sexting in public discourse from a playful pastime into an often-bizarre public spectacle.

Public Discourse on Sexting: From Humor to Moral Panic

Efforts to monitor public discourse on sexting often produce a puzzling mosaic of lighthearted and serious themes, however necessary they are to establish common understanding and consent. Humor about sexting is rampant in public discourse, although sexting is indeed a phenomenon with serious consequences for many participants. Even serious academic papers have titles such as "Sexting: Risky or [F]risky? An Examination of the Current and Future Legal Treatment of Sexting in the United States" (Shah, 2010) and "I'll Show You Mine, If You Show Me Yours: Public Policy Implications of Adolescent Sexting" (Soster & Drenten, 2011). A growing number of officials in public settings, athletes in sports arenas, employees in the workplace, and students in the classroom have been determined by the public or by themselves as engaging in sexting, often with very negative results for those labeled as involved in these practices (Kingston, 2009). Sexting can be unsettling to careers, community statuses, and family relationships, often in a way that other kinds of sexually related media exchange does not. Former U.S. Representative Chris Lee's digital photographs were widely disseminated in both humorous and serious media settings. *The Wall Street Journal* stated that Anthony Weiner's "lack of self-discipline" in sexting made him vulnerable to attack: "He has shattered his ability to serve their [his constituents'] interests. In an age of aggressive computer hacking, he also put himself at risk of blackmail by criminals or adversaries" ("Anthony Weiner's Judgment," 2011, p. A16). Association with high-profile cases has made discourse on sexting issues less nuanced and more sensationalistic. The "legitimate" applications of computing technology to sexual interaction (such as the corporate dating website eHarmony.com) are often contrasted sharply with means considered less wholesome (such as Craigslist's personal sections described in Tossell, 2009, often associated with prostitution).

Production of mediated sexual content does not require technological support; drawings on cave walls can depict sexual activity. However, the kinds of interactive exchanges supported by the Internet involve fairly advanced technologies. In past decades, the VCR and related video and music recording devices served to displace a good deal of the control of the movie and music industries over distribution (Benson-Allott, 2007; McLeod, 2005; Oravec, 1996). For example, the VCR created options for the independent recording and playback of sexual expression. Cases of piracy of copyrighted materials were often construed with the emergency rhetoric of a "moral panic" (Patry, 2009), with disputes over market rights being recast with the rhetorics of emergency and crisis. Söderberg (2010) provided comparable accounts of how "hacking" has been characterized. These cases provide a number of useful parallels to sexting. Many young people were saddled with criminal complaints and civil judgments for downloading music and bypassing the commercial music channels; "remix culture" and related perspectives were developed in defense of these activities (Lessig, 2004).

There has indeed been a great deal of commercial news coverage of sexting issues: since many people are curious and want to learn about information and communications technologies, coupling descriptions of innovations with storylines involving villains and victims provides opportunities for broadcast and print media to attract reader attention. However, since those who engage in pirating, hacking, and sexting are generally not producing a commercial product or service, defense of these individuals in the press has been rather muted. Sexting, for example, involves the noncommercial production and exchange of sexual goods by individuals and is rarely given the support that deep-pocketed sexual purveyors (such as sexual magazine publishers) can and do obtain. Individuals snapping a photo of what is inside their underwear are producing a sexual good, but they are likely to be attacked in the public sphere in a way that marketers who regularly fill billboards in public arenas with sexual content are not.

Part of the more legitimate aspects of the enhanced public concern related to sexting relates to the age levels of individuals participating in these practices. They are reportedly decreasing, with individuals as young as 9 years old allegedly involved (Doherty, 2010). An Australian study recommended that cyber-safety education should start "as early as kindergarten" to curtail these practices (Dickinson, 2011).

Sext Education

Individuals monitoring information about sexting may indeed tap the resources being produced in educational contexts. Awareness of sexting trends has already affected some of the potential interactions between and among teachers and students; many of the well-publicized cases of sexting involve students and teachers (or coaches). In a move linked to such narratives, the State of Missouri passed legislation in 2011 that restricts social media contact between teachers and students. For example, teachers cannot "friend" students on Facebook. However, other sorts of efforts have been more empowering to educators in these contexts and even interesting and edifying to their audiences as well. At the very least, sexting-related instruction includes some information about new social media technologies, about which many individuals are curious, and combines it with discussions of sexuality.

Efforts to control the sexual behavior of others have long consumed the attention of legal authorities, educators, social workers, and other concerned individuals (Bashford & Strange, 2004). Those who want to learn about sexting or to warn individuals against it have a variety of cautionary productions to utilize in their efforts, including fictional presentations and narratives. Emerging figures relating to sexting include the frantic parents who are attempting to undo the damage done to their children by sexting incidents. Therese Fowler reflected on her son's 2009 real-world sexting incident as described here:

> Fowler's then-19-year-old son was arrested on a misdemeanor charge for e-mailing nude photos of himself to a 16-year-old female friend. "I was astonished that simply sharing a photo of himself with a girl he knew could be considered criminal," says Fowler, 44, from her home in Wake Forest, N.C. (Donahue, 2011, p. 44)

A set of "mental hygiene" or "social guidance" films attempts to shape student lifestyles by presenting cautionary scenarios for consumption by students in high school classrooms (Galloway, 2010). The genre of the mental hygiene film has long roots in social engineering efforts; previous titles include *Boys Beware* (1961), warning young men of the reported dangers of encounters with male homosexuals (McCain, 1999). Several DVDs have been developed with such titles as *The Dangers of Texting and Sexting: What Kind of Message Are You Sending?* and *Texting and Sexting: Think before You Hit Send.* The former portrays a girl who "sends a provocative photo of herself to a boy against her friend's advice, and the photo is passed around" (Gallego, 2011, p. 52); the latter describes a "girl upset to discover that a locker room photo of her had been taken without her knowledge and was being sent from phone to phone" (Reutter, 2011, p. 52). The 2010 DVD *Photograph,* produced by CentaCare Sandhurst Loddon Mallee Cyber Safety Project, "Developing Ethical Digital Citizens," and funded by the Telstra Foundation, attempts to demonstrate the "repercussions when a fifteen-year-old girl sends an inappropriate photograph of herself to her boyfriend" ("*Photograph:* A New DVD-ROM," 2011, p. 61).

The DVD entitled *The Realities of Sexting (You Can't Unsend!)* is part of the Social Sensibilities Series offered by Learning Seed, an instructional video producer in the United States (Bilmes, 2011). It portrays sexting in gender-specific and often sensationalistic terms, highlighting dangers for young women:

> [It] begins with a dramatization of a girl sexting a photo of herself and its disastrous consequences. Viewers are given advice on how to flirt and fit in without having to resort to sexting. High school students comment throughout, and a diverse cast of teens are featured in dramatizations. A high school media literacy instructor offers occasional advice, but most of this powerful presentation consists of the narrator spelling out the dire consequences of sexting. (Bilmes, 2011, p. 54)

According to their producers, these videos are generally intended to provide contexts for the discussion of sexting issues in classrooms and community centers in efforts to frame sexting issues for young people and parents. Approaches vary; Hicks (2009) emphasized "coolness" in strategies in the following terms: "If we want to manage their [teens'] 'sexting' behavior, we have to convince them that truly cool people don't do this" (p. A16). Other approaches incorporate more intense shaming efforts, emphasizing the stigma involving release of certain personal images or narratives. Others emphasize bullying as a theme: sexting has indeed been a considerable aspect of a number of bullying or sexual harassment incidents, especially in contexts involving young people. It has been linked with a number of shocking and saddening events involving minors; for example, in 2009, Hope Witsell (a 13-year-old) committed suicide after digital images of her breasts were distributed in an assortment of high school venues (Szymialis, 2010).

Other forms of educational interventions dealing with sexting are increasing in number and variety, including the course "Cell Phone Photography" (taught by Sean Flannery at Immaculata University in Pennsylvania), which is described as addressing the dangers of sexting. Methods of discussing sexuality in university contexts have aroused concern, such as the demonstration of how to reach orgasm with a sex toy reportedly provided in a class on human sexuality and mass media at Northwestern University (Sergeant, 2011). Discussion of the criminal aspects of sexting is often incorporated into educational and public service efforts. Criminalization of sexting in Western nations has been problematic (Forbes, 2011); various legal scholars have labeled the images as "self-produced child pornography" (Hoffman, 2011, p. 1). Bowker and Sullivan (2010) discussed the "proposed blanket decriminalization of juvenile sexting in states including Vermont, Ohio, and Utah" (p. 27). Some U.S. states have worked to moderate the kinds of treatments given to minors who are involved in sexting practices:

> In Nebraska, it is now legally permissible for persons under the age of 19 to have possession of a sexually explicit visual depiction of another youth who is at least 15 years old if that youth voluntarily shared the image and the image was not further disseminated. Alternatively, a youth who *further* distributes that image to others may face criminal felony charges (Nebraska Legislative Bill 97, 2009). Similarly, the governor in Illinois signed a bill into law in 2010 reducing penalties for consensual sexting to a misdemeanor, with penalties increasing for further distribution, up to a felony charge for distribution on the Internet. Additionally, the law creates noncriminal penalties in the forms of monetary fines, diversion programs, and loss of access to electronic communication devices (Illinois Public Act 096–1087, 2011). (Segool & Crespi, 2011, p. 30.)

Until such decriminalization or at least moderation efforts are widespread (and successful), monitoring of the societal weather conditions related to sexting is needed for responsible interaction in this arena.

Some Conclusions and Reflections

From development of online dating websites to the sending of a single photograph from a mobile phone or personal computer, individuals are incorporating digital images into many sexual contexts along with related narrative. This has been labeled as "sexting," and it is likely to continue well into the future. Sexting practices are evolving, constructing a new set of social compacts concerning how the artifacts and traces associated with sexual activity are created and handled. Important ethical issues are emerging involving the uncertainties of sexting which require considerable amounts of information about recent societal trends and technical context to interpret. Efforts to glean this information require nontrivial allocations of time and attention. Some of these issues involve the malleability of the digital images involved in sexting as well as their rapid and sometimes enigmatic dissemination on the Internet; this potential for modification makes whatever eventually transpires in terms of consent problematic. The sensationalizing of these matters in public discourse has often transformed relatively harmless and mundane communications activities into potentially damaging ones: the phenomenon of sexting has conjoined the many uncertainties involved with the Internet and social media with apprehensions about sexuality. Individuals need to make considerable efforts in order to understand the societal and technical contexts of sexting sufficiently so as to obtain the levels and forms of consent needed to be ethically responsible. Not long ago, the expression of profane words involving sexuality was coupled with a considerable level of analysis and subsequent requests for consent. Comedians such as Lenny Bruce and George Carlin suffered arrest for uttering certain sexually charged words in public and many individuals would ask permission of their listeners (or at least provide a warning) before using certain terms referring to sexuality. Transmissions of sexually themed images today are treated in comparable ways in many contexts; those involved in these practices are called upon to understand the societal context and afford participants at least some sort of notice before sexting is conducted.

Participation of young people in sexting practices compounds the anxieties involved. It also provides an easy target for those who want to attack individuals who seldom have the legal means to fight back. Piracy scandals involving Napster and other networks of the past decade underscore how moral panics can wreak havoc on the lives of individuals. People who produce sext messages in non-commercial contexts are routinely being attacked although the media environment in many Western nations is itself saturated with images and text involving sexuality. Individuals must navigate a confusing array of advertisements, billboards, popular music productions, movies, and television shows designed to initiate sexual response, with young people often being especially targeted (Levande, 2008; Nosko, Wood, & Desmarais, 2007). This response is then associated with the purchase of a particular item or service. Sext messaging, however, is often met with negative consequences. Gender-related dimensions of these issues are especially problematic: much of the discourse on how young women are affected by sexting behavior (either as senders or receivers) places them in passive roles (Egan & Hawkes, 2008). The discourse often helps to fuel a moral panic rather than equip all parties involved to find authentic and empowering expression online. Penny (2011) described the UK's response to sexting as a "state-sponsored panic." The voyeuristic dimensions of a good deal of media coverage of these issues are obvious as reflected in the cases chosen to highlight and the descriptions of the plights of the parties involved.

Discussion of these issues has varied from extraordinarily serious, as in Darden's (2009) suggestions to school administrators for how to handle the aftermath of suicides linked to sexting, to trivial and entertaining. The characterization of Congressman Weiner's sexting incidents as "Peter Tweeters" by Jay Leno on *The Tonight Show* is an example of the latter (Leno, 2011). The casting of individuals in sexting scenarios in starkly negative roles, as victim and perpetrator, can do little to empower individuals to use online communications effectively and humanely. Individuals have some responsibilities in the face of moral panics to protect each other as well as themselves, whether the panic concerns the file-sharing system Napster or sexting. Individuals also have responsibilities to work to improve how matters involving personal reputation and well-being are construed, especially when people in fragile job markets and severe economic conditions are involved. We need a nonsensationalistic "safe sexting" initiative that provides enhanced digital citizenship for those who treat each other's communications with respect.

It is indeed doubtful that the quantity of sexual interactions portrayed on the Internet and related communications channels will ever be reduced. The rush and "naughtiness" of sexting may certainly be part of its attraction (as is its coupling with automobile driving), and fusing it with interesting and involving social media platforms only intensifies its appeal. Those who design educational agendas or public policy approaches that attempt to curtail sexting can see their efforts backfire as they produce sensationalistic materials related to sexting. By sending sext messages with pictures of identifiable bodily parts, individuals relinquish control over aspects of their self-presentation. Participating in something that is ill-advised and makes one vulnerable may not in itself be problematic in an ethical sense. However, risking the well-being of another individual, especially in a context in which "sexting" is regarded in such negative terms, is yet another matter. An action as trivial as hitting the "reply all" button rather than the "reply" can effectively distribute intimate materials to vast and unknown audiences.

Sexting issues will continue to present themselves in various forms, as new forms of technology-enhanced sexuality and sexual expression emerge. Many kinds of sexual communication and interaction are moving to online systems, including those incorporating a substantial component of risk. The character of these interactions is unfortunately being shaped by the rhetorics of emergency and crisis. Well-tempered and informed approaches toward these issues are likely to produce the most useful results in terms of policy and practice.

References

Anthony Weiner's judgment. (2011, June 7). *The Wall Street Journal—Eastern Edition,* p. A16.

Bashford, A., & Strange, C. (2004). Public pedagogy: Sex education and mass communication in the mid-twentieth century. *Journal of the History of Sexuality, 13*(1), 71–99.

Benson-Allott, C. (2007). VCR autopsy. *Journal of Visual Culture, 6*(2), 175–181. doi:10.1177/1470412907078558

Bilmes, D. (2011). The realities of sexting (you can't unsend!). *School Library Journal, 57*(2), 54–55.

Bowker, A., & Sullivan, M. (2010). Sexting: Risky actions and overreactions. *FBI Law Enforcement Bulletin, 79*(7), 27.

Boyle, S. (2011, August 18). Mankini, the latest word in male bathing garments and the *Oxford English Dictionary. The Mirror,* http://www.highbeam.com/doc/1G1–264538511.html

Custódio, L. (2011). *They still matter! The importance of print and analog media for youth civic participation in the digital age.* IAMCR OCS, IAMCR , Istanbul. http://iamcr2011istanbul.com/

Darden, E. C. (2009). When tragedy strikes. *American School Board Journal, 196*(9), 50–51.

Dickinson, A. (2011, June 22). Sexting laid bare: Call for earlier education. *The Courier Mail (Australia),* p. 4.

Doherty, E. (2010, February 11). Nine-year-olds "sexting." *Herald Sun (Australia),* p. 21.

Donahue, D. (2011, April 28). Fowler gives "exposure" to consequences of sexting. *USA Today,* p. 4d.

Egan, R. R., & Hawkes, G. (2008). Endangered girls and incendiary objects: Unpacking the discourse on sexualization. *Sexuality & Culture, 12*(4), 291–311. doi:10.1007/s12119–008–9036–8

Eraker, E. C. (2010). Stemming sexting: Sensible legal approaches to teenagers' exchange of self-produced pornography. *Berkeley Technology Law Journal, 25*(1), 555–596.

Forbes, S. (2011). Sex, cells, and SORNA: Applying sex offender registration laws to sexting cases. *William & Mary Law Review, 52*(5), 1717–1746.

Fowler, T. (2011). *Exposure.* New York, NY: Ballantine.

Gallego, B. (2011). The dangers of texting and sexting: What kind of message are you sending? *School Library Journal, 57*(5), 52.

Galloway, L. (2010). Thumbs down: An anti-"sexting" video ends up victim-blaming instead. *Bitch Magazine: Feminist Response to Pop Culture, 48,* 10.

Gibbs, N. (2009). Cell-phone second thoughts. *Time, 173*(10), 56.

Godson, S. (2010, May 8). Is "sexting" my husband a good idea?; Sex counsel. *The Times* (London), p. Weekend10.

Goodson, M. (2011, August 24). Keeping teens from hitting "send." Schools crack down on indecent photos, messages on cellphones. *Dallas Morning News,* p. 1A.

Harrison, M. A. (2011). College students' prevalence and perceptions of text messaging while driving. *Accident Analysis & Prevention, 43*(4), 1516–1520. doi:10.1016/j.aap.2011.03.003

Hewitt, B., & Driscoll, A. (2009). The dangers of 'sexting.' *People, 71*(12), 111–112.

Hicks, J. (2011, August 18). NJ schools prepare to implement bullying law. *The Philadelphia Inquirer,* p. B01.

Hicks, M. (2009, April 15). "Sexting" uncool, but also wrong. *The Washington Times,* p. A16.
Hoffman, J. (2011, March 27). States struggle with minors' sexting. *The New York Times,* p. 1.
Juntunen, L., & Valiverronen, E. (2010). Politics of sexting. *Journalism Studies, 11*(6), 817–831.
Kingston, A. (2009). The sexting scare. *Maclean's, 122*(9), 52.
Leno, J. (2011, August 16). *The Tonight Show* [Television series]. Burbank, CA: NBC.
Lentini, R. (2011, March 31). Sexting for "insurance." I'll show you mine, if. . . . *Herald Sun* (Australia), p. 16.
Lessig, L. (2004). *Free culture: How big media uses technology and the law to lock down culture and control creativity.* New York, NY: Penguin Press.
Levande, M. (2008). Women, pop music, and pornography. *Meridians: Feminism, Race, Transnationalism, 8*(1), 293–321.
McCain, R. S. (1999, December 9). New book unreels unreal "mental hygiene" films; Students of '50s, '60s cautioned on dating, sex, drugs. *The Washington Times,* p. A2.
McLeod, K. (2005). MP3s are killing home taping: The rise of Internet distribution and its challenge to the major label music monopoly. *Popular Music & Society, 28*(4), 521–531. doi:10.1080/03007760500159062
Miller v. Skumanick, 605 F. Supp. 2d 634 (M.D. Pa. 2009).
Nosko, A., Wood, E., & Desmarais, S. (2007). Unsolicited online sexual material: What affects our attitudes and likelihood to search for more? *Canadian Journal of Human Sexuality, 16*(1/2), 1–10.
Oravec, J. (1996). *Virtual individuals, virtual groups: Human dimensions of groupware and computer networking.* New York, NY: Cambridge University Press.
Oravec, J. (1998). Every picture tells a story: Digital video and photography issues in business ethics classrooms. *Teaching Business Ethics, 3*(3), 269–282. doi:10.1023/A:1009869905418
O'Riordan, A. (2010, April 4). Sexting not harmless fun for the cheated wife. *The Sunday Independent* (Ireland), p. x. Here is the online link: http://www.independent.ie/opinion/analysis/sexting-not-harmless-fun-for-the-cheated-wife-2124670.html
Patry, W. (2009). *Moral panics and the copyright wars.* New York, NY: Oxford University Press.
Paul, P. (2011, July 17). He sexts, she sexts more, report says. *The New York Times,* p. 6.
Penny, L. (2011). There's more to the Facebook generation than the odd poke. *New Statesman, 140*(5060), 12.
Phone ban in school changing rooms to stop "sexting." (2011, February 10). *The New Zealand Herald.*
Photograph: A new DVD-ROM about "sexting." (2011). *Screen Education* (61), 77.
Reutter, V. (2011). Texting and sexting: Think before you hit send. *School Library Journal, 57*(5), 52.
Risling, G., & Jablon, R. (2011, October 12). Hollywood hacking case victims include Johansson. *Boston Globe,* p. A1.
Robinson, M., & Picard, D. (2009). *The framed world: Tourism, tourists and photography.* London, UK: Ashgate.
Segool, N. K., & Crespi, T. D. (2011). Sexting in the schoolyard. *Communique, 39*(8), 30–31.
Sergeant, M. (2011). Let's talk about sex. *Psychologist, 24*(5), 336.
Shah, K. (2010). Sexting: Risky or [f]risky? An examination of the current and future legal

treatment of sexting in the United States. *Faulkner Law Review, 2*(1), 193–216.

Simpson, M. (2011). *Rights watch—presumed guilty: The harrowing tale of how one NEA member's investigation of student sexting led to his own prosecution on child pornography charges.* New York, NY: National Education Association.

Söderberg, J. (2010). Misuser inventions and the invention of the misuser: Hackers, crackers and filesharers. *Science as Culture, 19*(2), 151–179. doi:10.1080/09505430903168177

Soster, R. L., & Drenten, J. M. (2011). I'll show you mine, if you show me yours: Public policy implications of adolescent sexting. *AMA Winter Educators' Conference Proceedings, 22,* 216–217.

Stokes, P. G. (2010). Young people as digital natives: Protection, perpetration and regulation. *Children's Geographies, 8*(3), 319–323. doi:10.1080/14733285.2010.494891

Sullum, J. (2009). "Sexting" scare. *Reason, 41*(1), 16–17.

Szymialis, J. J. (2010). Sexting: A response to prosecuting those growing up with a growing trend. *Indiana Law Review, 44*(2), 301–339.

Tossell, I. (2009, May 1). Sex, murder and the outbreak of moral panic. *The Globe and Mail* (Canada), p. R21.

Wiederhold, B. K. (2011). Should adult sexting be considered for the DSM? *CyberPsychology, Behavior & Social Networking, 14*(9), 481.

Wysocki, D., & Childers, C. (2011). "Let my fingers do the talking:" Sexting and infidelity in cyberspace. *Sexuality & Culture, 15*(3), 217–239. doi:10.1007/s12119-011-9091-4

CHAPTER NINE

Griefing Through the Virtual World

The Moral Status of Griefing

ROLAND WOJAK

Introduction

Communication technology has evolved from sending letters to interacting with people across the world near instantaneously within virtual environments. The disparity is staggering. We use computers to talk in real time over the phone, via text messages, e-mails, forums, video calls, and virtual worlds. People use the Internet to learn and discuss, with classes being held online and papers being transmitted, read, and graded digitally. Students who may otherwise have had trouble succeeding in traditional language classes have the opportunity to use virtual media to assist them in learning a new language (Roed, 2003).

Compute-mediated communication has enabled us to communicate tremendous amounts of information over great distances in near real time, and it has gone even further in the form of computer-mediated environments. Through the use of computer-mediated environments, people are able to not only communicate but interact on a higher level through virtual actions within virtual environments. By creating and inhabiting a virtual representation of themselves, a person is able to traverse a virtual world and interact with other people within the world, even overcome physical disabilities to not only regain a sense of freedom and mobility but surpass what they once had by flying (Ford, 2001).

As our understanding of the human mind increases, we are finding the divide between the virtual and real beginning to blur. Clark and Chalmers have proposed that the mind can be extended outside of the body, that our cognitive system can include things like notebooks, calculators, and computer files (1998). Researchers are finding that nonverbal social behavior carries over into virtual worlds (Yee, Bailenson, Urbanek, Chang, & Merget, 2007), making them an inexpensive place to study diverse social interactions, such as cross-cultural and crisis management situations (Nakanishi, 2004). Virtual environments are even being used to study psychological disorders (Kim et al., 2009) and to treat autism (Altschuler, 2008; Turner, 2008).

The fact that virtual environments elicit the same kind of behavior found in face to face interactions indicates the profound similarities between what could be considered very distinct. These similarities are what enable us to treat autistic children via virtual interactions, but they are also leading people to reject the line that has traditionally divided reality and virtual reality. Jos de Mul claimed that, "in telepresence and virtual reality the artificial body has become part of our own body scheme," which could explain why "pilots exercising in such simulators often experience dissociation between their biological and artificial bodies," leading to imbalance and flight restrictions by some airlines after simulator use (2003, p. 259). Humberto Maturana Romesín claimed that, "the distinction between virtual and non-virtual realities does not apply to the operation of the nervous system" and that virtual realities have now become nonvirtual (2008, p. 109).

The benefits of this technology are numerous indeed, but the dangers are often overlooked and need to be considered. If things continue at this rate, interaction through computer mediated environments will only become more prevalent. What many people do not seem to realize are the real-world repercussions that result from events in virtual worlds, ranging from marriage to murder. As this aspect of human life continues to increase, it is important to look at both how we behave, and perhaps more importantly, how we should behave within virtual worlds.

Griefing in Virtual Worlds

One widespread behavior that deserves attention is what is referred to as griefing. As its name suggests, griefing is behavior that is directed toward eliciting grief in another human being. The goal of this chapter is to try and determine whether we can consider griefing as having a moral status, and if so, what that status would be. I hope to make it clear that in general, actions carried out within virtual environments are subject to moral consideration, and that in particular, griefing is wrong.

Before we begin discussion of the moral status of griefing, we must first understand exactly what griefing is. Griefing has been defined as the "intentional harassment of other players . . . which utilizes aspects of the game structure or physics in unintended ways to cause distress for other players" (Warner & Raiter, 2005, p. 47). This definition is rather basic, but it can serve as a template that can be applied to most games. However, it can, and probably should, be expanded to cover a wider range of actions. Chesney, Coyne, Logan, and Madden have defined griefing as the "intentional, persistent, unacceptable behavior which disrupts a resident's ability to enjoy Second Life and First Life" (2009, p. 542). There are two features of this definition that should be drawn out. The first is the mention of Second Life. Second Life is a virtual world where people can create avatars allowing them to interact with other users. More than just a game, Second Life can be seen as just that, a second life that users live through a virtual world. Within Second Life, companies have stores and nations have embassies, people fall in love and get married while others act as terrorists (http://secondlife.com). While the study focused on interactions among people within Second Life, the definition could be used to cover other virtual worlds as well. Second, and most importantly, the definition extends beyond Second Life to the effects the behavior has on people within First Life.[1] The effects of this behavior do not stop at the monitor but continue beyond, affecting the psychological and emotional well-being of a person in the real world. And it is there that we get to the true heart of the matter. It is little wonder that this behavior has earned the name "griefing," as it is a behavior intended to cause grief or suffering in another human being through actions taken in a virtual world.

Griefing as a Form of Cyberbullying

One way to look at griefing is as another form of cyberbullying. People have begun to extend the traditional view of bullying, namely, "abusive relationships where there are repeated, intentional hurtful actions directed against a victim (or victims) who is in a less powerful situation and thus not able to defend themselves" (Smith, 2009, p. 180), in order to encompass similar actions carried out through various technological media. For example, bullying carried out over the phone, through text messages, video recordings, web pages, posting of videos on web pages, and so forth, are all considered cyberbullying. When we look at the nature of griefing, it should be clear that griefing can be a subset of cyberbullying (owing to the fact that it takes place through the use of computer technology) and by extension, bullying in general.

Griefing is an intentional action that is meant to cause suffering in another human being, and, because most people do not choose to be victims of this sort of thing, the action is often directed at individuals who are unable to adequately defend themselves. Oftentimes griefing behavior can occur within an imbalance of power, and while the imbalance of power may not be physical, it is still used to inflict harm. Sometimes the imbalance of power may consist in the strength or abilities of the user's avatar, potentially allowing the user to kill another avatar in a matter of seconds (as is the case in games like World of Warcraft). Other times the imbalance of power may consist in knowledge of the virtual world. In cases like these, new players are particularly vulnerable as they tend to lack the knowledge to protect themselves. In one case, a new player was attacked in Second Life by four people who tossed her around and destroyed her house. They were only able to do this because at the time she did not know how she could protect herself (Chesney et al., 2009). Because of the similarities that exist between bullying, cyberbullying, and griefing, namely, that they are intentional actions meant to inflict harm against another person who tends to be in a lower state of power, people like Chesney et al. (2009) consider griefing to be "framed within the general bullying domain and can be considered as an extension to the phenomenon of cyberbullying" (p. 542).

To be certain, there are real differences between bullying and cyberbullying, and even more between the two and griefing, most notably the fact that traditional bullying takes place on a physical level whereas cyberbullying and griefing do not. There is also supposedly a greater possibility of escape from cyberbullying and griefing than there is from traditional bullying, at least in some cases. It may be harder to run from a bully in the school yard than to shut off your phone or computer, but it may be harder to escape from the effects of a webpage than from a bully in the school yard. A third difference is that oftentimes a bully's identity is unknown to the victim in cases of cyberbullying, as opposed to traditional bullying where the bully's identity can be quite clear. Often a victim will be bullied via phone or text messages from unknown or blocked numbers, and anyone can post videos or make a webpage with relative high degree of anonymity; the real-world identities of both victims and bullies may be unknown to each other in the case of griefing, however. Since virtual worlds by nature tend to obscure a user's real-world identity, it is often the case that people only know each other by their in-game identities, which, in some cases, can be easily changed by inhabiting another character.

Despite the differences, there are still strong similarities between all forms of bullying. Bullying at its root is an action designed to cause harm, in one form or another, in another human being. In traditional bullying, it is generally achieved through the physical power of an individual or group. But this is not always the case; sometimes the power lies in a person's social status or rank in some form of hierarchy. This abuse of one's rank is often seen in office bullying. The imbalance may also lie in a difference of technological knowledge, as is most often the case in cyberbullying and, sometimes, griefing. People who know how to hide their identities and make webpages are able to bully people online more proficiently than those who do not. Sometimes the knowledge concerns how a particular game works, resulting in new players being targeted for griefing as they lack the knowledge to defend themselves. Still, griefing may also closely resemble the imbalance of power found in traditional bullying. While not a difference in physical power of the individuals, there can be great differences in the virtual power of their avatars. This imbalance can be exploited in order to grief another player. Regardless of what form the imbalance takes, it is still an imbalance that is wielded with intent to harm another individual, generally over a sustained period of time. This holds true regardless of what form the bullying takes.

Bullying is widely considered wrong and is the subject of many studies and programs aimed at understanding and preventing the problem (Rivers & Noret, 2010; Slonje & Smith, 2008; Smith, 2009; Smith et al., 2008; Ybarra & Mitchell, 2004). It does not matter what form the bullying takes; there are still real-world consequences and real-world harm that is caused. Suicide is not an unheard of result, and when a child is lost, it does not seem to matter if the bullying was done online or in the classroom. One would be hard pressed to find an ethical theory that would not condemn such actions; why and how could we make an exemption for griefing?

Justifying Griefing

One possible motivation for griefing found in the Chesney et al. (2009) study on griefing in the virtual world Second Life was that for some users, Second Life was viewed as being just a game. As such, they either carried over the mentality of other games, where the goal is to kill or destroy, or they dismissed any complaints of griefing as unfounded because it was all taking place in a game and not in a real-life situation. While the motivation was originally classified as a single motivator, it seems that there are really two different ways to understand the claim that "it is just a game," and, from which two different possible ways to excuse the behavior from moral consideration can be drawn. The first is the mentality that this game is like many others, where the goal is to kill or destroy opponents. The second is that Second Life, like all other virtual worlds, is just a game, and as such, it is somehow detached completely, or at least removed to an extent, from reality in some significant way. It is this proposed detachment that somehow removes the moral status of actions performed within the virtual game environments.

When examining the actions of people within a virtual world, one of the first questions that should be answered is what the nature of that particular virtual world is. There are many different kinds of virtual worlds; indeed, there are as many virtual worlds as there are individual games and environments. There are worlds that are interacted with via gaming consoles, others by computers. Some worlds are constituted by one playable character controlled by one person at a time, and others are filled with thousands. The rules that form the structure of virtual worlds, which often happen to be what draws in individuals, vary from extremely structured to only minimally structured. Examples of the former are easily seen in many console games, where a player progresses through levels in a linear fashion. Examples of the latter can be seen in sandbox style MMOs (Massive Multiplayer Online games), where there are basic rules that determine what a player can do with a character but not much else to determine how players interact with each other within the world and even with the world itself.

One possible motivation for griefing behavior is that people who are accustomed to playing in worlds that are very structured and designed for players to kill or destroy either NPCs (nonplayable characters, or basically, computer-controlled characters) or other players, carry over that mentality into virtual worlds that are much less structured and present environments for people to interact with each other in more positive ways, such as Second Life.[2] In worlds such as Second Life that seem to truly enable people to live a second life, whether it be building relationships, learning, working for real-life money,[3] or simply having fun, the question of what kind of world it is and what rules, if any, govern the interactions within it needs to be answered. If, as in the case with many virtual worlds, the world is generally considered a game, then, like in other games, killing another player in the game world may not only be acceptable, but encouraged. If this is the case, then griefing should not be considered a morally wrong behavior but more like being tackled in football or shot in a FPS[4] game. While people may be able to make an argument for differentiating virtual worlds like Second Life from virtual worlds considered to be primarily game worlds in order to avoid this objection, they would not seem to be able to do the same for the many other virtual worlds which are widely considered to be games, even if their argument were successful. As such, there appear to be worlds where griefing behavior may not only be acceptable but encouraged based on the game's design.

The second possible interpretation of the "it's just a game" motivation for griefing is that all virtual worlds should be considered games, or at least nothing more than a fantasy world, and since the behavior takes place within these confines, it is not subject to moral evaluation.[5] From this motivation, an objection to evaluating the behavior of interaction within virtual worlds on a moral level can be established. The weak form of the objection would be that interactions within games are subject to different rules than those that govern behavior in general, and the strong form is that all interactions within virtual worlds are subject to different rules than those in the real world.

A clear example of how actions that take place within the confines of a game are not subject to the same rules that govern actions in the world in general can be seen in the unfortunate incident where Ray Chapman was struck in the head by a pitch, resulting in his death. After learning of his death, the pitcher, Carl Mays, turned himself in to the New York district attorney, who ruled it an accident. Despite throwing a pitch that caused the death of another person in front of thousands of witnesses, the rules of the game effectively exonerated him of any wrongdoing.[6] In a similar way, it can be argued that since virtual worlds are also games, as such the actions that take place within them are not subject to the same rules that govern actions in general. In this particular case, griefing should not be considered morally wrong, because it takes place in a game world.

A stronger claim is that, beyond whether or not we classify any particular virtual world as a game, the behavior of individuals within these worlds is not subject to the same rules that govern the behavior of people in the real world; since, of course, these actions do not take place in the real world but in a virtual world. I find this position more difficult to motivate. In order for this position to be held in good faith, there would need to be some important difference between interacting with people face-to-face and interacting with them through virtual media. That is not to say, of course, that there are not obvious, and real, differences between the two, but that we must determine whether these differences warrant attributing a different moral status to the same or similar behavior.

The most obvious difference is the lack of, or even potential for, physical contact. You can kill or hug another person's avatar in a virtual world, but this will cause no immediate physical sensation in either person. Some may see the lack of any physical damage as a significant enough reason to dismiss claims of any wrongdoing. Another major difference is the ability to escape any unwanted behavior. It is true that people are able to do this in face-to-face interactions, to some extent, but it is much easier to walk away from the computer than to escape a bully in the school yard. So, because it is easy to escape from the behavior, there is no need to even worry about whether or not it is wrong.

There are, of course, situations where these differences are significant enough to warrant a change in the moral status of an action; killing a person in real life, for example, is quite different from killing another player's avatar in a virtual world. The issue at hand, however, is whether these differences are enough to justify changing the way we morally evaluate griefing, and, as I hope to make clear, they are not.

The Moral Status of Griefing

Claiming that griefing is alright because it is "just a game" may be an excuse, but it is not a motivation for the action. The motivation seems to be something very much akin to traditional bullying. This can be seen from the fact that griefing necessarily involves one person intentionally causing suffering in another through the virtual environment. If killing in a game was all that it was about, then the griefer could play other games where this was the primary objective of all the players (single or multiplayer) or they could PvP (player versus player combat) within the rules of the game. The fact that they do neither of these shows that they are griefing for other reasons; generally, this involves whatever reasons a person may have for causing suffering in another.

While some may feel the status of griefing comes down to how we classify certain virtual worlds, that is, as games or virtual worlds, this distinction is irrelevant to the moral status of griefing. First, the nature of the behavior is not in line with the goals of, at least, most games (Warner & Raiter, 2005). Second, regardless of where the action takes place, it is nonetheless an action committed by an individual and, in this case, it is an action that is directed toward another individual; and, since the effects that result are not limited to the virtual world but are indeed quite real, the actions themselves have moral status.

The first way to interpret the claim that griefing is morally permissible because it is all "just a game" is to see griefing as part of the game. In this sense, people who are accustomed to playing games where the expressed goal is to kill or destroy are bringing this mindset to other games. The problem is that these other games are not the same games that are designed around killing or destroying. There is a big difference between first person shooter (FPS) games and most MMOs and an even bigger difference between FPS games and virtual worlds like Second Life. In FPS games, there is a clearly defined goal, that is, to kill your opponents. And while this goal may have certain variations, from killing as many as possible to killing them in order to secure or defend a position, or some other objective, the main goal remains the same. When someone brings this same mindset into MMOs or virtual worlds like Second Life, where the "goal" is, to some extent, nonexistent, the individual is not making it alright to grief other players; he or she is just mistaken. Bringing a certain mentality to a virtual world is not enough to excuse bad behavior; just think how absurd this would be if practiced in real-world interactions. Just because I like kickboxing in the ring does not mean it is alright for me to drop a roundhouse on some stranger's head in the supermarket.

If bringing a certain mentality to a game or virtual world is not enough to excuse griefing, then perhaps an excuse can be found within the rules of the game. It is clear that there are some games where killing is not only condoned but encouraged. There are many FPS games where players actively seek competition with one another. The game mechanics of most MMOs and virtual worlds also allow for PvP, as such, the rules of those games allow for the killing of other players. Since the rules of games change the way we morally evaluate actions (think of Ray Chapman and Carl

Mays), games that allow for the killing of other players (or any other action) thereby excuse the behavior from moral consideration. This argument rests on one major assumption that needs to be sorted out, that is, if an action is possible within a game or virtual world, then it falls under the rules of the game.

While game and virtual worlds usually do not come with a rule book like games played out in the real world, they do come with some rules, generally in the form of a EULA (end user license agreement) or a terms of service agreement. In addition to containing regulations concerning what a player is able to do with the program itself, there are also rules concerning a player's conduct in the game world. These rules are enforced by GMs[7] and generally govern a player's conduct in regard to NPCs and other players.

The existence of explicit rules, and entities charged with enforcing them, offers insight into the matters at hand; specifically, it shows that not every action that is possible in the virtual world thereby falls under the rules of the game. Not only are there explicit rules concerning behavior, but there are individuals whose job it is to enforce these rules. If any action that was possible were thereby allowed under the rules of the game, there would be no need for either. This leaves the argument that griefing is permissible because it is possible within the virtual world untenable.

It is now clear that just because an action is possible does not mean it is allowed by the rules of the game, but it would still be possible to argue that griefing would be exempt from moral judgment if it could be shown that it was not prohibited by the rules of the game. The first step would be to look through the EULA (which most games require each time a player logs into the virtual world) to see if there is any clause that prohibits griefing. There are many games that offer players a large degree of freedom and ample amounts of PvP, but one of the most brutal is a game called Darkfall.[8] Darkfall is a virtual world left largely open for players to decide what to do. Inhabited by various difficult monsters, the game is made even more dangerous by the fact that you can kill and be killed by anyone at any time; furthermore, upon death, you are returned to a specific location chosen by you in the past (potentially hours away from where you died), while everything you had on your person or in your bag stays at your gravestone, where you died, for anyone to loot. Players are able to band together for protection and even able to own hamlets and cities and the resources they provide. These holdings are not permanent, however, as they can be sieged and taken away by other groups of players. In such a brutal environment, people would assume that griefing would not only be rampant but that it would not be prohibited within the EULA; they would, however, be mistaken. Within Darkfall's EULA Version 1.0.48, there are numerous clauses prohibiting what is commonly considered griefing, the most notable of which are the first two under the player conduct subsection within the in-game and in-world conduct section:

i. While playing Darkfall online, you must respect the rights of others and their rights to play and enjoy the game.
ii. You may not defraud, harass, threaten, or cause distress and/or unwanted attention to other players within the Game, the World or on the official Darkfall online web sites (https://accounts-us.darkfallonline.com/account/tos.php).

By comparison, the terms of service agreement for Second Life requires adherence to a set of community standards called the "Big Six":

1. Intolerance: Combating intolerance is a cornerstone of Second Life's Community Standards. Actions that marginalize, belittle, or defame individuals or groups inhibit the satisfying exchange of ideas and diminish the Second Life community as a whole. The use of derogatory or demeaning language or images in reference to another Resident's race, ethnicity, gender, religion, or sexual orientation is never allowed in Second Life.
2. Harassment: Given the myriad capabilities of Second Life, harassment can take many forms. Communicating or behaving in a manner which is offensively coarse, intimidating or threatening, constitutes unwelcome sexual advances or requests for sexual favors, or is otherwise likely to cause annoyance or alarm is Harassment.
3. Assault: Most areas in Second Life are identified as Safe. Assault in Second Life means: shooting, pushing, or shoving another Resident in a Safe Area (see Global Standards below); creating or using scripted objects which singularly or persistently target another resident in a manner which prevents their enjoyment of Second Life.
4. Disclosure: Residents are entitled to a reasonable level of privacy with regard to their Second Life experience. Sharing personal information about your fellow Residents without their consent—including gender, religion, age, marital status, race, sexual preference, alternate account names, and real-world location beyond what is provided by them in their Resident profile—is not allowed. Remotely monitoring conversations in Second Life, posting conversation logs, or sharing conversation logs without the participants' consent are all prohibited.
5. Adult Regions, Groups, and Listings: Second Life is an adult community, but "Adult" content, activity and communication are not permitted on the Second Life "mainland." Such material is permitted on

> private regions or on the Adult Continent, Zindra. In either case, any Adult content, activity, or communication, that falls under our Adult Maturity Definition must be on regions designated as "Adult," and will be filtered from non-verified accounts. Other regions may be designated as either "Moderate" or "General." For more information on how to designate land, events, groups, and classified listings, please carefully read the "Maturity Definitions."
>
> 6. Disturbing the Peace: Every Resident has a right to live their Second Life. Disrupting scheduled events, repeated transmission of undesired advertising content, and the use of repetitive sounds, following or self-spawning items, or other objects that intentionally slow server performance or inhibit another Resident's ability to enjoy Second Life are examples of Disturbing the Peace (http://secondlife.com/corporate/cs.php).

These two very different virtual environments share basic common rules that govern user behavior. Second Life is a virtual world where people can come together and interact within a virtual environment and is defined as such in the terms of the service agreement (http://secondlife.com/corporate/tos.php#tos4), not as a game (though there are games that can be played within the virtual world). Darkfall, on the other hand, is a brutal game consisting of a virtual world where players can interact through a virtual environment, most often in order to kill another player and take everything he or she is carrying. Despite the great differences between the two virtual worlds, both explicitly protect a user's rights to play and enjoy the world and to be free from harassment. Both implicitly protect users from griefing.

While there may be virtual worlds that do not prohibit griefing, if a brutal game like Darkfall does, it is probably a safe bet that most others do as well. Still, there is a possibility that griefing would not be prohibited by a virtual world's EULA, raising the question of how to deal with that world.

Even if griefing is not prohibited by the EULA of a particular virtual world, it is still not exempt from moral consideration. In such worlds, we can look to the game design for hints to the intended purpose of PvP or try and view it in light of the perceived spirit of the game; regardless of how we approach it, it seems like the burden of proof should not be on the person condemning the behavior but those trying to condone it. Just like arguments are needed to allow for murder to be morally acceptable under certain conditions, so to must there be arguments that allow for griefing. Without these arguments, it makes no difference whether the EULA prohibits the action, since we have already shown that just because an action is possible does not mean it is permissible.

Of course, there is always the possibility that griefing would be explicitly allowed, in which case it would fall under the rules of the game. Just as punching is not allowed except in certain situations, in a boxing ring for example, griefing could have a place in certain virtual worlds, though one would be hard pressed to find one.

There is a final argument that could exempt griefing from moral consideration. If we understand the claim that griefing is permissible because the virtual world is "just a game," to mean that all virtual worlds, regardless of classification, are somehow separate from the real world in such a way that it exempts actions performed within them from moral consideration, then griefing would no longer be subject to moral consideration. While this position would be very strong if it is capable of being supported, it is much more difficult to do so than it may at first appear.

No one is going to deny that there are differences between virtual actions and those that take place in the "real world" (most notably the lack of physical consequence and the ability to escape more easily), but while these differences do change how we evaluate some behavior, they do not change the fact that we do evaluate the behavior. We do not hold someone responsible for murder if they kill another player's avatar, as no murder has actually occurred, but we do, or at least should, hold someone morally responsible for continually killing a person's avatar so that the individual is unable to do anything but die. And why should this not be the case? While there may not be physical consequences, like a bruise or a broken bone, actions performed in virtual worlds do, nonetheless, have real-world consequences. Even if the interaction takes place through a virtual medium, it is still an interaction between people. People build meaningful relationships, both in game and out, that can even lead to marriage. Others have killed themselves (Kuhn, 2009) or others (Chesney et al., 2009) over events in virtual worlds. There can be no denying that there is a significant connection between actions in virtual worlds and real-world effects. According to Coyne, Chesney, Logan, and Madden (2009), over 40% of those surveyed believed griefing to have at least the same, if not more, of an impact on victims than traditional bullying, with the targets of griefing rating it as being more harmful than those who had not been griefed. The harmful effects of griefing may not be physical, but they are no less real.

Even if there are such negative consequences, there is no obligation for people to play these games, because unlike real life, they are able to shut off the computer and exit the virtual world, thereby escaping from any possible harm. While true, this should in no way change how we judge the actions themselves. We do not want to say that it is alright if I bully someone because they are able to go home and escape me if they want. Bullying behavior is generally considered wrong, whether it be in person, online, by phone, text messages, or e-mails, all of which offer various ways to escape. Why should we consider griefing, cyberbullying through virtual worlds, to be exempt from moral consideration when we do not consider these other forms exempt? Just as we would not want to excuse bullying over the phone or through instant messages, we should not excuse griefing just because it takes place through a different medium.

But is it not possible that people who are negatively affected by griefing are only affected so because they have an unnatural attachment to their avatar? Current research into the nature of virtual reality and the connection people have to it, however, does not support the claim that all psychological harm experienced by a person is due to an unnatural attachment to his or her avatar. In fact, it appears that a person's behavior can be influenced both in and out of the virtual world by something as simple as an avatar's appearance (Yee, Bailenson, & Ducheneaut, 2009). Players normally refer to themselves in the first person when playing and talking about actions that they have performed within virtual worlds (Cogburn & Silcox, 2009). Romesín, after showing how the nervous system does not operationally distinguish between virtual and nonvirtual realities, claimed that "virtual realities are never trivial, because we always become transformed as we live them . . . regardless of whether we like it or not" (2008, p. 109). In light of this, it would seem more unnatural for a person to lack an attachment to an avatar than to be affected by virtual actions experienced through it.

Even if a person were to withdraw attachment to an avatar, it would not change anything about how we judge the behavior itself. Griefing is an action directed at another human being with the intention to cause suffering in that person. It does not matter that it is done through a virtual world or how much a person feels attached to his or her avatar. Avatars are not NPCs; they are the virtual manifestation of a person meant to represent that person within a virtual world, and no amount of attachment or lack of attachment to the avatar can change the fact that griefing is an action that is directed at the user. When a person spawn camps another person, so that person can do nothing but die as soon as he or she returns to his or her corpse, this does not frustrate the avatar but the user controlling it. Just as we would not want to excuse the murder of a person hopelessly lost to nihilism who genuinely does not have any attachment to his or her own life, we should not excuse griefing because someone feels attached or unattached to his or her avatar. As has been argued by Jessica Wolfendale (2007), if the attachment to an avatar is morally insignificant and should be done away with because it can lead to suffering, then we would need to say that this should be done for all attachments, as there does not appear to be any good reasons why avatar attachment should be different from property, family, friend, community, ideals, or any other form of attachment.

It also is not a matter of knowing the real-world identities of the people involved. How absurd would this criterion be if we used it for all other ethical concerns? Imagine a world where people were only held accountable for wrongs committed against those they knew and not strangers. The bottom line is that actions performed within virtual worlds are really still actions performed in the real world by a real person directed at another person through a particular medium with real-world consequences and effects.

We are now in a position to ask what the moral status of griefing is in relation to some of the major ethical theories. If, as I have hoped to make clear, the actions that take place in virtual worlds are encompassed by what we consider to be the real world, then, apart from any rules of a particular game, the actions that take place within virtual worlds should be held to the same standards as the actions that take place in common face-to-face interactions. As Ashley John Craft put it, "users have the same *de facto* duties towards each other when they interact within virtual spaces as they do when writing in print, talking over the telephone, or meeting in person"; they are, after all, still interacting with another human being (2007, p. 216). There are good reasons to believe that griefing in a virtual environment would result in moral condemnation by three of the major ethical theories: Kantian deontological ethics, utilitarian consequentialism, and virtue ethics.

Basically, Kantian deontological ethics calls for people to respect each other and treat one another as ends in themselves. Accordingly, we should treat others as we would like to be treated. There is, of course, much more to this theory, but there is no need to go into greater detail here. It should be clear that when people grief others, they are not treating them as an end in themselves but as a mere means for their own enjoyment. Furthermore, there are good reasons to believe that one should refrain from similar behavior in regard to actions in virtual worlds that do not even involve other people. According to Philip Brey (1999), there are two arguments that support this claim: the "argument from moral development" and the "argument from psychological harm." The first claims that performing harmful acts against virtual entities can lead a person to treat other people in the same way. The second claims that people may suffer psychologically when "representations of themselves or individuals like them, or representations of other beings, or things that they value, are not treated with respect by others" (p. 9). While the second argument may be more controversial, there is evidence to support the argument from moral development (Porter & Starcevic, 2007; Vessey & Lee, 2000).

The same sorts of arguments could be used to support a utilitarian consequentialist moral evaluation of the behavior. According to utilitarianism, an action should be performed only if it results in maximizing utility (basically, the greatest amount of good for the greatest amount of people). Again, it would seem clear that players who engage in griefing behavior are not maximizing the good as they harm not only those they grief but potentially themselves (according to virtue ethics, as I discuss next) and other people out in the world. If there is a link between playing violent games or engaging in unethical behavior within virtual worlds and a greater disposition to do the same in face-to-face interactions with other people (as seems to be the case), then behaving in this way very well might result in a much greater net negative utility. Regardless of how other people may be affected, there is always the way a behavior affects the agent.

The goal of virtue ethics is eudaimonia, or human flourishing, generally achieved by practicing virtuous actions and thereby developing a virtuous moral character. One central aspect of the theory concerns the motives for the behavior, that is, that the action is done for the right reasons and does not just happen to coincide with what a virtuous person would do. Another important aspect is that a virtuous character is built from habit, in other words, virtues become character traits. If by the habituation of certain actions a person builds his or her character, then it does not seem hard to make the case that acting immorally in virtual worlds would lead to the development of a poor moral character. As Thomas Nys (2010) so eloquently put it, "by playing such games one grows callus in one's soul" (p. 85).

Conclusion

Just as any other action that takes place in the real world is judged from the viewpoint of one particular ethical system or another, so too should actions that take place in virtual worlds (as they are really just extensions of the real world anyway). There is not, however, always straightforward agreement between theories or between interpretations of the same theory, even regarding the same situation. Nonetheless, there are good reasons to believe that most, if not all, ethical theories would consider griefing to be wrong to some degree. It would appear at first glance that griefing would be considered morally reprehensible according to Kantian deontological ethics, utilitarian consequentialism, and virtue ethics. But before any moral status can be judged, it must first be determined whether a particular action is subject to moral consideration. Therefore, there is a question that is fundamental to all the theories; namely, whether we are even capable of considering actions in virtual worlds morally.

I hope by now that it has been clearly shown that despite the differences between the virtual and real worlds, the actions performed in the former are still subject to moral consideration, just as those performed in the latter; it is all occurring in the same world, after all. There are real-world consequences for actions in the virtual world that factor into the real-world moral status of actions. There is, of course, the possibility that a game will be created that makes it clear that the goal is to grief other players, but until such a time, we cannot appeal to the fact that griefing can take place in game worlds in order to shield the action from moral consideration. The rules of most, if not all, virtual worlds in existence today do not condone griefing and even prohibit it. Griefing is not part of the game; it is an abuse of game mechanics, allowing players to subvert the spirit of the game in order to inflict suffering on another human being. Whether or not any particular ethical theory considers this morally wrong is not in any way changed by the fact that the actions occur within a game or virtual world. And while I have not been able to go through every theory under every interpretation, I think it is safe to assume that most would consider intentionally causing suffering in another human being to be wrong.[9]

Notes

1. First Life is the term many users in Second Life use to refer to what most would call the real world as opposed to the virtual world.
2. This position is exemplified by the response of one of the participants in Chesney et al.'s (2009) study. Participant (VK) said, "I think some people think of SL like other games such as World of Warcraft and think the idea is to destroy everything" (p. 539).
3. Linden Dollars can be exchanged for real-world currency and vice versa.
4. First person shooter; this term describes a game structure where a player inhabits a character through a first person camera perspective and engages in armed combat with other players, generally using firearms.
5. A clear example of this position is seen in the response by (JE), another participant in Chesney et al.'s (2009) study. (JE) claimed that, "as much as we would like to play 'make believe' and pretend that real people are under attack. . . . It is just a virtual world, real rules can and do not apply" (p. 540).
6. This is a well-known historical fact to some, but for others a description of it and how it relates to games and justice can be found in Greg Lastowka's (2010) book, *Virtual Justice* (p. 102).
7. Game masters are people hired by the company to monitor actions within the virtual world, help players with issues, and punish players when necessary (including both temporary and permanent bans).
8. See http://www.darkfallonline.com.
9. While it may seem clear that griefing is wrong, the degree of wrongness (if there is such a thing) remains open. Similar to how we would say that all bullying is wrong but some incidents display a sort of wrongness well beyond simple school yard taunts, there are times when we would want to say that a particular griefing event is somehow worse than others. Most likely what makes one particular act of griefing worse than another depends largely on the context. Since most of the context (particularly things about the people involved) is hidden in most online interactions, perhaps the best thing to do would be to avoid the behavior altogether.

References

Altschuler, E. L. (2008). Play with online virtual pets as a method to improve mirror neuron and real world functioning in autistic children. *Medical Hypotheses, 70,* 748–749.

Brey, P. (1999). The ethics of representation and action in virtual reality. *Ethics and Information Technology, 1,* 5–14.

Chesney, T., Coyne, I., Logan, B., & Madden, N. (2009). Griefing in virtual worlds: Causes, casualities and coping strategies. *Info Systems Journal, 19,* 525–548.

Clark, A., & Chalmers, D. (1998). The extended mind. *Analysis, 58*(1), 7–19.

Cogburn, J., & Silcox, M. (2009). *Philosophy through video games.* New York, NY: Routledge.

Coyne, I., Chesney, T., Logan, B., & Madden, N. (2009). Griefing in a virtual community: An exploratory survey of Second Life residents. *Journal of Psychology, 217*(4), 214–221.

Craft, A. J. (2007). Sin in cyber-eden: Understanding the metaphysics and morals of virtual worlds. *Ethics and Information Technology, 9,* 205–217.

Ford, P. J. (2001). Paralysis lost: Impacts of virtual worlds on those with paralysis. *Social Theory, 27*(4), 661–680.

Kim, E., Jung, Y.-C., Ku, J., Kim, J.-J., Lee, H., Kim, S. Y., . . . Cho, H.-S. (2009). Reduced activation in the mirror neuron system during a virtual social cognition task in euthymic bipolar disorder. *Progress in Neuro-Psychopharmacology & Biological Psychiatry, 33,* 1409–1416.

Kuhn, D. (2009, February 19). *Addicted: Suicide over Everquest?* Retrieved from http://www.cbsnews.com/stories/2002/10/17/48hours/main525965.shtml

Lastowka, G. (2010). *Virtual justice: The new laws of online worlds.* London, UK: Yale University Press.

Mul, J. d. (2003). Digitally mediated (dis)embodiment: Plessner's concept of excentric positionality explained by cyborgs. *Information, Communication & Society, 6*(2), 247–266.

Nakanishi, H. (2004). Free walk: A social interaction platform for group behaviour in a virtual space. *International Journal of Human-Computer Studies, 60,* 421–454.

Nys, T. (2010). Virtual ethics. *Ethical Perspectives, 1,* 79–93.

Porter, G., & Starcevic, V. (2007). Are violent video games harmful? *Australasian Psychiatry, 15*(5), 422–426.

Rivers, I., & Noret, N. (2010). Participant roles in bullying behavior and their association with thoughts of ending one's life. *Crisis, 3,* 143–148.

Roed, J. (2003). Language learner behaviour in a virtual environment. *Computer Assisted Language Learner, 16*(2–3), 155–172.

Romesín, H. M. (2008). The biological foundations of virtual realities and their implications for human existence. *Constructivist Foundations, 3*(2), 109–114.

Slonje, R., & Smith, P. K. (2008). Cyberbullying: Another main type of bullying? *Scandinavian Journal of Psychology, 49,* 147–154.

Smith, P. K. (2009). Cyberbullying: Abusive relationships in cyberspace. *Journal of Psychology, 217*(4), 180–181.

Smith, P. K., Mahdavi, J., Carvalho, M., Fisher, S., Russell, S., & Tippett, N. (2008). Cyberbullying: Its nature and impact in secondary school pupils. *The Journal of Child Psychology and Psychiatry, 49*(4), 376–385.

Turner, M. (2008). Robots, mirror neurons, virtual reality and autism. *The Psychologist, 21*(8), 674–676.

Vessey, J. A., & Lee, J. E. (2000). Violent video games affecting our children. *Pediatric Nursing, 26*(6), 607–611.

Warner, D. E., & Raiter, M. (2005). Social context in massively-multiplayer online games (MMOG's): Ethical questions in shared space. *International Review of Information Ethics, 4,* 46–52.

Wolfendale, J. (2007). My avatar, myself: Virtual harm and attachment. *Ethics and Information Technology, 9,* 111–119.

Ybarra, M. L., & Mitchell, K. J. (2004). Online aggressor/targets, agressors, and targets: A comparison of associated youth characteristics. *Journal of Child Psychology and Psychiatry, 45*(7), 1308–1316.

Yee, N., Bailenson, J. N., & Ducheneaut, N. (2009). The Proteus effect: Implications of transformed digital self-representation on online and offline behavior. *Communication Research, 36*(2), 285–312.

Yee, N., Bailenson, J. N., Urbanek, M., Chang, F., & Merget, D. (2007). The unbearable likeness of being digital: The persistence of nonverbal social norms in online virtual environments. *CyberPsychology & Behavior, 10*(1), 115–121.

CHAPTER TEN

Permissible Piracy?

BRIAN CAREY

Introduction

Most people have reasonably well-developed moral frameworks which they use to determine how they ought to act in a wide variety of everyday situations. Much of this is cobbled together from our intuitions, perhaps reinforced here and there with more critical reasoning and self-reflection, especially when we find ourselves confronted with situations which cannot easily be accounted for by consulting our moral intuitions alone. It is possible that the speed of our technological advancements over the last several decades has been so quick that we have outpaced the development of our intuitive moral frameworks in some cases such that we find ourselves engaging with ethnical problems in areas where we have no clear intuitions to fall back on in trying to figure out how we ought to behave.

Digital objects present just such a problem. By "digital objects" I mean to refer to a broad class of items including music, television shows, movies, computer games, and other computer programs which are stored and transmitted electronically and differ from "physical" objects primarily in the sense that they may be replicated at (almost) no cost. This new ability to replicate property rather than remove it from the possession of its original owner creates a problem for our traditional understanding of morally impermissible theft which is predicated upon the assumption that every act of morally impermissible theft is an act of (nonconsensual) deprivation.

In attempting to address the issue of the illegal downloading of digital objects (digital piracy[1]), there are at least two perspectives one might take: the first is a legal perspective, focused on the question, "What sort of laws should we have?" with regard to digital piracy. This is not the approach I aim to take here. Rather, I shall be concerned with a second kind of approach, one which takes the perspective of the individual potential pirate and asks, "Under what circumstances, if any, is it morally permissible for me to illegally download digital objects?" The distinction between the two approaches is important because the principles which tell us what the law ought to be will not necessarily tell us how we, as individuals, ought to behave.

In order to answer this question, I shall begin with the basic premise that it is morally wrong to cause harm to another person without that person's consent in all circumstances, except when causing harm is necessary in order to prevent some greater harm from occurring.[2] I begin by outlining the standard argument most commonly used to dissuade digital pirates, which is based upon the (usually explicit) equation between digital piracy and theft. From here I move to an examination of theft more generally and conclude that we ought to think of theft as nonconsensual deprivation. Following this, I consider cases where digital piracy does not satisfy the deprivation condition, and I suggest that there are some cases where digital piracy may be permissible provided that the pirate would not otherwise have purchased the object in question.

Last, I consider the implications of this view and some of the additional ethical hurdles which must be surmounted before we may permissibly pirate. In response to the objection that the "theft as deprivation" view is too broad, I introduce a more nuanced view of deprivation whereby morally permissible deprivation is distinguished from morally impermissible deprivation on the basis that the former is necessary in order to prevent some greater harm. Here I also consider the method of downloading and suggest that moral pirates must ensure that they do not pirate in a way which is likely to facilitate morally impermissible theft by others (for example, via peer-to-peer file sharing). I leave it an open question as to whether these additional hurdles will simply make permissible piracy not worth the effort, but I shall conclude that there are at least some cases where an individual may participate in digital piracy without committing an act which is morally wrong.

You Wouldn't Steal a Car...

In 2004, the Motion Picture Association of America (MPAA) produced an antipiracy advertisement which typifies the most popular strategy for dissuading potential pirates: equating the pirating of digital objects, or intellectual property (in this case, digital copies of movies), with more "traditional" forms of theft. The 45-second advertisement opens with a shot of a woman sitting at a computer, about to download a pirated movie, before cutting away to a series of scenes involving a shadowy figure breaking into a car, stealing a handbag, a television, and so on. Each scene is accompanied with the words, "You wouldn't steal a car," "You wouldn't steal a handbag," and so forth, before ending with the words, "Downloading pirated films is stealing. Stealing is against the law. Piracy. It's a crime" (Loughlan, 2007, p. 401).

The advertisement's purpose is quite straightforward. Most people have strong intuitions when it comes to the morality of stealing a car or a handbag. If the equivocation between digital piracy and "traditional theft" can be made to stick, then people will be less likely to download pirated movies. The fact that an effort needs to be made to establish this connection, however, suggests that, at the very least, digital objects present certain unique conceptual quirks which make them distinct from physical objects like cars or handbags, and which make it difficult for many people to think of downloading a pirated movie as belonging to the same category of moral actions as shoplifting a physical copy of a DVD. The worst-case scenario, from the perspective of groups like the MPAA, is that these different types of action may in fact be *morally* distinct to the extent that, under certain circumstances, downloading a pirated movie should not be considered theft at all.

Theft as Nonconsensual Deprivation

Perhaps the most important distinction between physical and digital objects is encapsulated by the concept of "nonexclusivity." Edwin Hettinger (1989) described nonexclusive objects as follows:

> . . . they can be at many places at once and are not consumed by their use. The marginal cost of providing an intellectual object to an additional user is zero, and though there are communications costs, modern technologies can easily make an intellectual object unlimitedly available at a very low cost. (p. 34)

These features of digital objects introduce an obvious complication when it comes to how we conceptualize theft. If I steal your car, you no longer have the use of your car. But if I pirate your movie, you still have the original. Dictionary definitions of theft refer to "the taking and carrying away of the personal goods" (http://dictionary.reference.com/browse/theft) and the "taking and removing of personal property with intent to deprive the rightful owner of it" (http://www.merriam-webster.com/dictionary/theft). Legal definitions of physical theft[3] usually require some form of "appropriation" of property such that the thief deprives the rightful owner of his or her property. If what makes an act of theft depends on whether the owner of the property is deprived of its use (without the owner's consent), then it looks like downloading a copy of a movie doesn't satisfy this condition, since the original object remains in the possession of its original owner.

Fortunately, from the perspective of the MPAA at least, this definition of theft as "deprivation without consent" is far too narrow to be plausible. In order to see why this is so, we must recognize that the value of, for example, a car, includes not just its instrumental value as a vehicle to allow us to get from A to B. One's car (like virtually every possession one owns) also has potential value in terms of what it can be sold for at some point in the future. If I want to sell my car, but everyone who comes to view it can drive away in a copy of my car rather than pay the price I am asking, then I have been deprived of the income which would have been generated from the sale of the original. If we modify our understanding of deprivation to include deprivation not only of the original object itself but also of future monetary gain (such that deprivation of either of these is a sufficient condition for theft to have occurred), then we come closer to a more plausible definition of theft. If we apply this understanding of theft to digital piracy, then we can formulate a simple explanation as to why piracy is theft: when you download a digital copy of a movie, you are depriving the owners not of the movie itself but of the revenue which would have been generated had you purchased the movie through (legally) legitimate means.

The problem with this explanation, however, is that it assumes that every pirated copy of a movie translates into a deprivation of future revenue. It assumes that if the pirate were unable to download the movie, he or she would have purchased it. While this is undoubtedly true in many cases (perhaps the vast majority), it is certainly not true for all.

Why Do People Pirate?

Besides pirating digital objects merely as an alternative to buying them, there are a variety of other reasons why one might be motivated to download a digital object without paying for it, many of which should seem relatively benign in terms of the morality of the act. For example, some individuals will download copies of digital objects they have already paid for but which they want to back up on a secure location in case the original is lost or damaged. Others will download for the purpose of replacement—perhaps a disk has become scratched or a file corrupted and they can no longer access the content for which they have already paid. In some cases, it is simply more convenient for users to subvert copy-protection technology for products they have already purchased. For example, most video games can only be played on particular versions of particular operating systems, many require that a disc be inserted during play, and some require that an Internet connection be maintained at all times during play. Illegal software programs can often remove these inconveniences in ways which seem intuitively permissible.[4] If we were dealing with physical objects here, we might be tempted to say "tough luck" to those who have lost or damaged their original purchases. It would not be okay, for example, for me to steal a second copy of a book from a shop simply because my original copy was misplaced. But where the option exists to download a digital copy, we may well feel more sympathetic toward the downloader. Indeed, many artists and companies take these sorts of scenarios into account and consent to such practices. Some permit the downloading of digital materials provided that one has purchased the material in some form beforehand.[5]

Some pirates download on the basis that they wish to "try before they buy." Some video game pirates, for example, will frequently encourage downloaders to support the makers of the games by purchasing a legitimate copy, provided that they enjoy the game. In such cases, the availability of the piracy option may even generate more revenue as users who would not otherwise have bought a legitimate copy of the object in question are persuaded to do so by their positive experiences of the illegal copy. Given that some forms of piracy may in fact result in a net benefit to the rightful owners, the suggestion is sometimes made that some forms of piracy are justified because they actually help certain artists by propagating their material and allowing them to reach a wide audience of people who would otherwise not have been exposed to their work.[6] However, video game companies can and often do release demos of their products, just as musicians release sample songs and authors occasionally release sample chapters from their books online. Some well-established musicians can even afford to release entire albums without any fixed price, whereby customers pay as much or as little as they think the product is worth.[7] If the owners of the material judge that it would be in their best interests to release some or all of their work for free, they are in a position to do so. To make that choice for the owners by copying their work without their consent looks less like a sincere effort to help a business or career and more like an attempt to provide excuses for unethical behavior.

Nevertheless, property owners can only invoke the consent principle up to a point. There are well-established limits to property rights in extreme cases: you do not have the right to kick a drowning man off your boat simply because he did not ask permission before clambering aboard. Most pirates, however, can hardly protest that their desire to experience the latest Lady Gaga track is so fundamental that it overrides Lady Gaga's right to determine how her music is distributed. But this assumes that Lady Gaga has such a right in the first place. We can assume that she has the right not to be deprived of her music or the money generated through its sale, but this does not necessarily extend over all cases where someone might have a reason to download a copy of a Lady Gaga album. It is not difficult to construct a variety of hypothetical scenarios where a person might want to pirate a copy of an album (or a movie, or a video game, etc.) for one reason or another, but where that person would not have chosen to purchase the album if it were the case that piracy was not an option. In such cases, why should we think that we require the consent of the creator or owner of the property in order to use a copy of that property in a way which results in no immediate or future deprivation of any objects or money? In such cases, the creators/owners are apparently left no better or worse off regardless of whether we pirate their property.

The Counterfactual Condition

If what really makes theft morally wrong, and what artists and companies have property rights to protect against, is the harm caused by deprivation of goods or future profit, it would suggest that piracy is ethically permissible under a certain limited range of cases. These are cases where the counterfactual scenario in which the pirate did not have the option to illegally download would not have resulted in any additional revenue for the creator/owner of the property. If there is no deprivation in either scenario, there is no harm, and if there is no harm, piracy (in the minority of cases which satisfy the counterfactual condition) is permissible.

Consider some examples which would satisfy this condition. A person decides that he or she wants to play a particular video game, but that person is unable to find it for sale in any retail store or online, so he or she downloads a copy of the game. In this case, had the person not pirated the game, he or she would not have bought it (because he or she could not have bought it, as far as the person knew). Suppose that this person is then informed of a website where it is in fact possible to purchase the game and does so. Did he or she behave unethically at any point? Did he or she cause harm to anyone by downloading a copy of the game it was thought could not be obtained? It must be conceded, however, that most cases which satisfy the counterfactual condition are not likely to do so because piracy is the only way in which the product can be obtained. It's likely that the majority of cases will involve a lack of sufficient funds and/or enthusiasm for the product to motivate a person to purchase it.

One difficulty with this account is that, through experiencing the pirated product, people may change their mind as to its worth. Essentially, if they knew then what they know now, they would have been willing to pay the price rather than downloading it. At such a point, to continue to use the product seems clearly unfair toward the creator: suppose that your pirated copy of Lady Gaga's latest album (which you have come to adore) becomes corrupted and you have the option of purchasing a legitimate copy or downloading another pirated copy. In this case, it would surely be unfair to pirate a second time. Indeed, it is most likely unfair to continue to listen to the pirated copy once you reach the point at which you would be willing to pay for what you have experienced.

A second difficulty is that whether one would be willing to purchase a product often depends to a significant extent on the price. There are video games which I intend to purchase, once they go down in price after a period of several months, but which I would be unwilling to pay full price for at the moment, because I don't think I would be getting my money's worth. Surely I am obligated to wait until the price goes down rather than pirate it now. Here I think it's appropriate to bite the bullet and claim that I would be justified in pirating now but only provided that I also purchase a legitimate copy the moment it comes down to an affordable price. Of course, trying to figure out where exactly the line is to be drawn in such cases is going to be difficult and would provide an easy excuse to a pirate looking to salve his or her conscience. But in principle at least, no harm is being done to the retailer provided that it receives the money it would have received if downloading was not an option.

Deprivation Revisited

I began this discussion by identifying (nonconsensual) deprivation as the fundamental characteristic of acts of theft, which makes those actions morally wrong. If I own a car and you take it from me, you have harmed me because I have been deprived of the use of my car. If I own an original digital recording and you copy it rather than paying for it, you have harmed me because I have been deprived of the revenue you would otherwise have spent. If deprivation is what makes theft morally wrong, then we must conclude that the (minority of?) cases where piracy does not represent a deprivation of revenue are not morally wrong. There is a possible challenge to this view, however, which threatens to undercut the identification of the moral wrongness of acts of theft with nonconsensual deprivation.

Suppose that I enter a record store with the intention of buying the latest REM album, but upon my arrival I discover that Metallica has also released a new album. I consider my options and, having only enough money to purchase one album, I decide to buy the Metallica album rather than the REM album. On this account, it looks as though my actions, in choosing to change my mind, Metallica's actions in bringing out a new album, and the decision of the storekeeper to stock it, have all contributed toward REM being deprived of the revenue it would otherwise have made from my purchase of the group's album. Yet it would surely be absurd to conclude from this that REM has been stolen from.[8] If this is the case, then the original formulation of theft as nonconsensual deprivation needs to be either modified to avoid such absurd conclusions or abandoned altogether. Unsurprisingly, I shall opt for the former approach.

It cannot plausibly be denied that the aforementioned counterexample does contain cases of deprivation, and it cannot be denied that REM has been harmed (in the sense that the group has been left worse off) in being denied the money it would otherwise have received. In order to rescue the deprivation-as-theft account, we need to make a distinction between necessary and unnecessary deprivation, a distinction which can be derived from a more general distinction between necessary and unnecessary harm.

One of the most basic paradoxes one encounters when dealing with issues of rights and liberty in general is that in order to secure certain kinds of freedoms, other kinds of freedoms must be limited. The notion of territorial sovereignty, for example, can only make sense if it implies that some particular state has certain powers within its territory which other states do not have. Private property rights demand similar exclusions—if everyone has the same freedom to use my toothbrush, then it is not really *my* toothbrush in any meaningful sense. On an even more fundamental level, if I own my own body, I can only do so if I have special rights over its use, which are possessed by nobody else.

These moral and legal spheres of authority are not absolute; there are extreme cases where they may be overridden. A state may lose its moral claim to territorial sovereignty if it carries out ethnic cleansing against its citizens, my right to bodily integrity or personal liberty may be overridden if I assault you, and so on, but these exceptions only apply when they are necessary in order to protect some more fundamental right. This is why, for example, our intuitions may tell us that it is not really morally wrong to steal a loaf of bread to feed one's family, while it is morally wrong to steal a car because one wants to go for a joyride.

These spheres of authority are detrimental to those who are excluded in one sense—I would certainly be better off financially if I did not have to purchase all of the digital objects that I desire, but they are vital in order for society to function. Consider the sorts of contradictions that would arise if everyone were obligated to never cause any sort of deprivation to others: I would be obliged not only to buy the REM and Metallica albums but to purchase as many albums as I possibly could, as to do otherwise would be to deprive the owners of potential revenue. Of course, this now leaves me worse off, depriving me of income I would rather have spent elsewhere, which means I have been harmed, which means it was wrong for Metallica and REM to release their albums in the first place, which deprives their fans of listening to their music, which harms them, and so on ad infinitum. Under such a system, given that resources are not infinite, it would be virtually impossible for anyone to act without causing some kind of deprivation to someone else. The harm caused to an individual by being excluded from certain rights, therefore, is vastly outweighed by the benefit that an individual gains in being able to exercise those same powers of exclusion over his or her own possessions and in securing the sorts of freedoms we need to be able to engage in commercial transactions in general.

One way to respond to the counterexample, therefore, would be to deny that it is an example of nonconsensual deprivation. Any reasonable individual would consent to a system whereby consumers are free to choose to spend their money as they see fit, where this level of freedom is necessary in order for commercial transactions to take place. In the aforementioned case, both REM and Metallica would surely consent to a system where consumers are entitled to change their mind.

Even if such consent is not present, however, we can invoke a general consequentialist principle for having the sort of system that entitles a consumer to change his or her mind—it's simply the only sort of model which could possibly work if we want to function as a society. The deprivations which arise as a result cause a kind of harm, but this harm is a necessary, minimal, and inevitable by-product of the system as a whole. This distinguishes it from other forms of harm which constitute instances of unnecessary deprivation. It is not necessary for me to function as an autonomous agent for me to deprive you of your car so that I may go for a joyride, but it may be necessary for me to deprive you of a loaf of bread in order to feed my starving family, or, in the aforementioned case, it is necessary that I be able to decide to purchase one product rather than another or to purchase no products at all.

It might also be noted here that consequentialist arguments for certain property configurations are especially prevalent when it comes to intellectual property such as patents. Though the ability for one company to patent a product causes harm to potential competitors, patent law is often said to be justified because it provides an incentive for innovation and results in a net utility gain when compared to possible alternative configurations.

So, while it is the case that almost all nonconsensual deprivation causes harm, it is not the case that all forms of harm are morally impermissible. In some cases this is because of the presence of mutual consent. For example, a boxer is harmed when he gets punched in the face, but it is not normally unethical for his opponent to punch him. In other cases this is because the kind of deprivation in question is part of a system which is necessary to secure broader, more important, freedoms in general. Acts of theft are wrong when they cause unnecessary deprivation, but not all acts of deprivation are unnecessary and therefore do not amount to acts of (morally wrong) theft.

A Final Hurdle: Aiding and Abetting Unethical Piracy

Even if we accept that digital piracy is not unethical in the limited number of cases where the counterfactual condition is satisfied, there is at least one additional moral hurdle that remains before we may permissibly download a digital object. This concerns the actual method one uses to download the digital objects in question.

Consider, for example, BitTorrent, a popular peer-to-peer file sharing protocol developed in 2001 which is used to transfer large files between computers and, according to some estimations, accounts for somewhere between 27% and 55% of all Internet traffic as of 2009 (Schulze & Mochalski, 2009). Ordinarily, a computer downloads a file from a single source, a process which can place considerable strain on the host computer and network, especially if the file is large and multiple computers are attempting to download it from the same source. Using BitTorrent, however, the file is broken into hundreds or thousands of pieces, which are then distributed to the computers attempting to download the complete file. Once a computer has begun to receive pieces of the file, it is then able to simultaneously upload these to other computers while it downloads the remaining pieces. This process incorporates each computer attempting to download the file (the "leechers") and each computer that has downloaded a complete version of the file (the "seeders"). As the leechers finish collecting the different pieces needed to complete the file, they become seeders, all the while distributing the pieces they have collected to other computers in the network (or "swarm"). When it comes to large files (typically television programs, movies, or video games), this process is almost always far more efficient and convenient than the traditional one-to-one download.

These downloading methods are problematic because, rather than involving merely a one-to-one exchange between the host of the pirated file and the downloader, BitTorrent-style protocols create networks of users who are ethically as well as electronically enmeshed. In these cases, the downloader is simultaneously an uploader, hosting the file in whole or in part for others to pirate. This expanded role brings with it an expanded ethical landscape. By facilitating the downloading of a file by others who may not satisfy the counterfactual condition, the downloader/uploader is almost certainly facilitating theft. (It should be noted that these concerns also apply to individuals who offer pirated files via direct one-to-one download to anonymous users online.)

There are at least two ways one could try to defend the uploading of pirated material in this fashion. One approach would be to point to the sheer number of computers, of seeders and leechers that typically participate in the simultaneous uploading and downloading of files, and to argue that the addition or subtraction of one additional computer to the "swarm" would make such an imperceptible difference to the system as a whole as to be morally insignificant.

This sort of argument brings us to well-trod territory concerning collective moral action and responsibility. Rather than becoming sidetracked by those debates, however, it should be sufficient here to imagine the consequences of endorsing this sort of justification. If a single participant of the swarm can be absolved of moral responsibility on the basis that his or her actions are neither necessary nor sufficient to cause the theft of digital objects, then the same excuse can be deployed for every single member of the collective (perhaps with the exception of the original uploader of the file), and we are left with the absurd conclusion than none of the subsequent pirates can be held morally responsible for their actions simply because so many have acted unethically.

An alternative attempt to defend uploading would begin by drawing a distinction between providing the opportunity to commit a theft and committing the theft itself. To take a nondigital example, a seller of guns might protest that he simply provides tools to individuals, tools which may be used for ethical or unethical purposes, and that he is not responsible for how his customers choose to use those tools. If a man murders his wife with a hammer, we do not usually assign responsibility to the shopkeeper who sold him the weapon. Why should we treat uploaders any differently? Assuming that some kinds of downloading are ethical, they are merely providing people with the resources to download files. If some should choose to use these resources for unethical purposes, that is not the responsibility of the uploader. This approach is perhaps slightly more plausible than the first since it appeals to our intuitive support for the idea that individuals must be held responsible for their actions and should not be allowed "pass the buck" in a way which divests them of some degree of responsibility which falls instead upon the enablers of the agent's action rather than the agents themselves.

To return to the world of digital downloads, in a response to legal threats from Irish record labels in 2009, Ireland's largest ISP, Eircom, blocked access to a popular file-sharing website called ThePirateBay. Because ThePirateBay allowed people to access illegal copyrighted materials, and Eircom allowed its customers to access ThePirateBay, the implication was that Eircom could be held partially responsible for the illegal activities (though this assumption has not been tested in Irish courts). In response to the targeting of ThePirateBay, it was pointed out that popular search engines such as Google, Yahoo, and Bing also facilitate illegal activity by allowing users to search for illegal files.

The chief difference between a website like ThePirateBay and a website like Google (and the reason why one is targeted by record companies while the other is not) seems to be that the main purpose of ThePirateBay is to allow users to circumvent copyright, whereas the main purpose of Google is not.[9] There seems to be a kind of "benefit of the doubt" principle at play here: while Google *can* be used to break the law, that's not what it is intended to do and that's not what the majority of its users use it for, in contrast to ThePirateBay. This matches our everyday intuitions quite well: if a cashier sees that I want to purchase a ski mask, a pair of night vision goggles, and several dozen boxes of ammunition, we might well think that the person has a responsibility to at least call attention to my suspicious behaviour. On the other hand, while I could probably commit murder by bashing my unfortunate victim over the head with a pineapple, seeing one in my shopping basket should not raise the suspicions of the shopkeeper, and he or she should not be held responsible to any degree for whatever wicked deeds I might subsequently commit.

With the aforementioned in mind, we can return to our case of the uploader of pirated material. In this case, the uploader likely realizes that the vast majority of those who are downloading the material he or she is providing are doing so to avoid paying for it. If this is the case, the uploader is clearly deserving of some degree of blame for the actions of those he or she facilitates, and we cannot accept the excuse that the uploader is merely providing resources or tools for people to use as they see fit.

Conclusion

Are there any circumstances in which it would be morally permissible for one to illegally download digital objects? I have suggested that there are, provided that one would not otherwise (or could not otherwise) have bought the object in question and provided that one obtains the objects in a way that is not likely to facilitate people engaged in morally impermissible acts. Living up to these demands may well prove so costly that the cases that do actually fall into the category of permissible piracy are extremely few. There is clearly a danger, furthermore, that we might be tempted to be too generous when we examine the counterfactual scenarios in order to determine whether we would have otherwise purchased a particular product.

If these ethical hurdles can be surmounted, what follows from this in terms of how the law ought to regard such ethical pirates? Not a lot, as it happens; in practice there would be no way for a court to determine whether a pirate would otherwise have bought a particular product. From this we can make a clear and straightforward consequentialist case that the law should not take such claims into account at all. The alternative would be that no pirate could be prosecuted at all, as each could claim that he or she satisfied the counterfactual condition. So, even ethical pirates may still be punished, and the position I have argued for here says nothing one way or the other with regard to the legal status quo. Nevertheless, my position suggests that piracy may be ethical under certain circumstances, a position which, I hope, can go some way to explaining our apparently confused moral intuitions surrounding the relatively new phenomenon of digital objects.

Author's Note

The author wishes to express his gratitude to the editors of this volume, the organizers and participants of the Digital Ethics Symposium at Loyola University, Chicago, and to Jonathan Flynn and Nicola Mulkeen for their helpful comments on earlier drafts of this article. Correspondence concerning this chapter should be addressed to Brian Carey, Politics, University of Manchester, UK. E-mail: bpdcarey@gmail.com

Notes

1. While I use the familiar terms "pirates" and "piracy" throughout this chapter, it should be noted that these labels are not unproblematic. Suzannah Mirghani, for example, has argued that "anti-piracy discourse is being sounded through an increasingly militarized language that relies on metaphors of war to inspire fear among audiences and to criminalize even the most casual of informational exchanges. Current anti-piracy campaigns discursively frame copyright infringement as a dangerous crime by connecting it to the word 'piracy' and to its cut-throat predecessor, open-sea piracy, to hold back the tide of wide-spread proliferation" (2001, p. 113).
2. To defend a particular conception of harm or a particular system of ranking harms according to the aforementioned premise is beyond the scope of this chapter; however, the aforementioned (admittedly rough) formulation should be sufficient in this context.
3. See, for example, the definitions of "theft" in the Criminal Justice (Theft and Fraud Offences) Act, 2001, Part 2, Section 4.1 [Ireland, 2001], the Theft Act, 1968, Section 1 [UK], Title 18, U.S. Code—Chapter 31 [United States, 2006], and Criminal Code Part 9, Section 322 [Canada, 1985].
4. For more on these sorts of cases, see Lessig (2001, p. 187).
5. The video games retailer http://www.gog.com, for example, allows customers who have purchased a game via its store to download unlimited copies of their purchase. The digital distribution platform Steam allows customers to access video games from any computer provided that they log in with an authorized account.
6. Some video game retailers even encourage "piracy" of their software: the terms and conditions for the video game "Freespace 2" explicitly authorize customers to make copies for their friends and acquaintances, provided that this is done on a noncommercial basis.
7. Perhaps the most famous example of this is the album *In Rainbows* by Radiohead, which was released digitally in 2007 and allowed customers to pay as much or as little as they wished. Video game developers have also used this model effectively; the "Humble Indie Bundle" was a collection of several games by independent developers which were sold without digital rights management (DRM) and allowed customers to pay as much or as little as they wanted. Four such packages have been released as of October 2011, resulting in over $6 million in sales.
8. I am indebted to Ian Carroll for raising this objection.
9. There is, of course, a less optimistic possibility, whereby the only reason sites such as ThePirateBay are pursued while sites such as Google are not is merely because the former make for easier targets.

References

Criminal Code, R.S.C. 1985, c. C-46 § 322, [Canada]. Retrieved from http://laws-lois.justice.gc.ca/eng/acts/C-46/

Criminal Justice (Theft and Fraud Offences) Act (2001) [Ireland]. Retrieved from http://www.irishstatutebook.ie/2001/en/act/pub/0050/index.html

Hettinger, E. C. (1989). Justifying intellectual property. *Philosophy and Public Affairs, 18*(1), 31–52.

Lessig, L. (2001). *The future of ideas: The fate of the commons in a connected world.* New York, NY: Random House. Retrieved from http://lessig.org/blog/2008/01/the_future_of_ideas_is_now_fre_1.html

Loughlan, P. L. (2007). "You wouldn't steal a car": Intellectual property and the language of theft. *European Intellectual Property Review, 29*(10), 401–405.

Mirghani, S. (2001).The war on piracy: Analyzing the discursive battles of corporate and government-sponsored anti-piracy media campaigns. *Critical Studies in Media Communication, 28*(2), 113–134.

Schulze, H., & Mochalski, K. (2009). Internet study 2008/2009. Retrieved from http://www.ipoque.com/sites/default/files/mediafiles/documents/internet-study-2008–2009.pdf

Theft Act (1968) [United Kingdom]. Retrieved from http://www.legislation.gov.uk/ukpga/1968/60/contents

Title 18, U.S. Code, *§641–669,* 2006 edition, [United States]. Retrieved from http://www.law.cornell.edu/uscode/text/18/part-I/chapter-31

CHAPTER ELEVEN

Legionnaires of Chaos

"Anonymous" and Governmental Oversight of the Internet

ALEX GEKKER

Anonymous is infinity divided by 0. = Syntax error
—ENCYCLOPEDIA DRAMATICA, n.d., a.

Tracing Anonymous

As the Internet gained dominance as a part of the everyday physical world, so did the governmental oversight of the online sphere. Rather than a disconnected cyberspace of disembodied personas, as portrayed in early days, the Internet—and more so the Web—became a crucial component of commerce, governance, and media. Power structures around the world have responded to this growing importance by reducing the tolerable margins of "devious" Internet behavior. This is done via legislation and oversight implemented by governmental and regulatory structures. This chapter aims to describe this phenomenon through a particular case study of "Anonymous," a leaderless, shapeless, online gathering, which emerged from idiosyncratic web culture and became an actor in global politics. By using Actor-Network Theory (ANT) as a methodological framework, I examine Anonymous and the influence it exerts on user control online.

This chapter first discusses the affordances of ANT in understanding distributed networks like Anonymous. Then, I offer a brief discussion of hackers and their traditional role online, while suggesting that despite their public image, governments and corporations have in fact enjoyed a cordial relation with this subculture over the years. I then describe Anonymous first as a web collective and then as a quasipolitical organization. I show how, despite its reputation, Anonymous does not constitute hackers in the traditional sense of the word, and how this fact underlines its relations with the authorities. Later, I discuss how the convergences of the offline world with the online, together with the discussed unique characteristics of Anonymous, position it as a threat in the eyes of governments worldwide. My aim is to show how a multitude of varying factors has led to increased resistance to Anonymous **because** it is not composed of hackers and how it may, in fact, contribute to the limiting of user control online rather than empowering it.

Actor-Network Theory (ANT)

The ANT perspective (Latour, 1987, 2005; Law, 1992) argues that in order to understand modern society, a researcher must follow the work-nets[1] of human and nonhuman actors (or rather, actants) through cultural-material artifacts. We can thus facilitate meanings by tracing and relating the different actors one to another. One must discard the theoretical constructs (such as ideologies) that are invisible and thus irrelevant to the actors in the system. Thus it is possible to locate the underlying currents in the decision-making processes of specific endeavors and to learn about the constructions of symbiotic human-technological relations in society. This process, through which a multitude of heterogeneous participants and the links they create is reduced to a seemingly monolithic process or organization, is called "punctualization."

To analyze Anonymous from ANT methodology, we must take into account the sociological and anthropological perspectives on the origins of the organization; its places of gathering and methods of communication (which both rely, to a great extent, on technical means); the current interplay of commercial, political, and private actors that operate within the Web; as well as the specifics of recent changes in the attitudes toward cyberspace deviation from nation-states and corporations, as reflected in lobbying and the legislation following it.

Another methodological note must be made here. ANT theory has originated in STS and organizational studies. The methodology was developed in laboratories, plants, and offices, with strong anthropological overtones. It suggests participant observation as one of the main tools or at least the ability to interview the actors in question in order to trace their action. It is more of a sociological "field" study that requires direct interaction with the objects of inquiry rather than an office analysis of data. Anonymous is problematic in the sense that you have nothing to question, no field to enter for inquiry. As is discussed further on, the group lacks formal representatives and membership. Furthermore, as recent inner chat logs of the Anonymous operation have disclosed (Cook & Chen, 2011), the members have a fondness for purposely misleading anyone who tries to gather insights from their gathering.[2] In my research, I followed a methodology described by Roversi (2008) in his study of hate groups of the Net: inside observation of the openly available sections while applying analysis based on other sources external to the group.

Hacker Culture

Anonymous is not composed of hackers. At least, that's how group members think of themselves[3] or the image they try to project. While a detailed account of what Anonymous is follows, we must also consider what it is not. It is easy to pin the characteristics of this collective as based in "Hacker Culture" (Levy, 1984; more on this later), but that categorization alone is insufficient for understanding Anonymous' motivations, objectives, or the way it challenges concurrent power structures. In Latourian terms, hacker culture is an intermediary rather than a mediator—a collective term for motivations and practices of quite a large and diverse group.

The Internet always had a place for libertarian individuals who used their technical skills to bend and break rules. Sociologist Manuel Castells (2001) argued that hackers are one of the pillars of modern web culture. Unlimited access to information, disdain for authority, and the desire to prove intellectual capability are the paramount ideals that drive this unique subculture. Yet hacker culture originated before the Web. Some of the Internet's most popular applications, from e-mail to the Web itself, were created by individuals following their curiosity and working on personal technological projects rather than on what they were supposed to be doing (Barabási, 2002; Castells, 2001). The culture is about discovery and innovation albeit not in the formal way. Technology researcher and critic Howard Rheingold, a central figure in one of the first countercultural digital bulletin boards, WELL ("Whole Earth 'Lectronic Link," which predated the World Wide Web by several years) goes as far as to claim that those libertarian values are imbued in the technological understructure of the Internet. He quoted his WELL colleague and the founder of the Electronic Frontiers Foundation, John Gilmore, and explained that, "The net interprets censorship as damage, and routes around it" (Rheingold, 2000, p. xxii). While technically inaccurate (censorship is difficult, but possible, as some authoritarian states prove time and time again), these attitudes show how much the hacker culture is perceived to be rooted in libertarian ideologies and why many consider those same principles to be imbued in the very foundations of modern online life.

Governments and corporations have tolerated this culture, because despite its informality and the tendency for insubordination, those talented tinkerers have generated real value. Castells (2001) suggested that the predominant hacker culture that was spread in U.S. and U.K. universities and similar facilities was largely responsible for the Western technological advantage in the Cold War. Soviet researchers, despite scientific excellence, were too stuck in political oversight and efficiency plans to commence the sporadic breakthroughs in computer sciences and electronics that characterized the West. Hackers were instrumental to the construction of the network society (Van Dijk, 2005) by providing the technological distributed networks that allowed Computer Mediated Communications (CMC) to substitute face-to-face communications in some personal and business aspects, while bringing some of their own free-spirited culture into those networks. Yet they were never countercultural in the literal sense of the word.

Hackers are by definition more interested in the development and spread of technology rather than in its social context. By observing modern-day Silicon Valley giants, which originated in the early days of the Net, one can see how corporations were started, alliances were formed, and positions of power accepted. As anthropologists Gabriella Coleman and Alex Golub (2008) argued, hacker culture should be viewed as a constellation of shifting rhetoric forms, each representing certain attitudes toward political notions of liberalism and centered on a different understanding of a specific technology's beneficial role in society. Although not mentioning ANT by name, their argument can be seen as influenced by its underlying concepts, as they stress conceptualizing hackers in accordance with their technological preferences, which suggest underlying social attitudes. Hackers, they say, "move in and between various ethical positions" (p. 271) while experimenting with new technologies, copyright types, and the like. This flux often positions hackers as a centerpiece of the technological establishment: CEOs, entrepreneurs, and investors, undermining the contra-cultural mantle which this group supposedly opposes. As Wayner (2000) noted in his book on free software, hacker culture is primarily *technological*, rather than *ideological.* One does not stop being a hacker by starting a corporate job or getting elected to a governmental position. Notorious hackers can become board members of ICANN (International Corporation for Assigned Names and Numbers, the principal governing body of the Internet; Castells, 2001, p. 32) or become chief scientists, designing the next generation of the Internet for the U.S. government (Lanier, n.d.).

As an illustration, one can consider the distinction between "white hat" (hackers working for legitimate security organizations) and "black hat" (hackers involved in borderline legal activity, often considered harmful) activities. Those two practices, both attributed to hacker culture, carry within themselves opposite values, while mostly using the same set of technologies and skills. The terms may describe a certain hacker not according to his or her ideologies but in relation to the particular goals that he or she pursues while employing hacking at a given moment. A security researcher may use his or her skills for a "white hat" day job, researching and defending against security threats in a large organization, while utilizing the same skills to conduct cyber attacks against ideological or personal targets after hours as a "black hat" hacker. In fact, it is not unheard of for a caught "criminal" hacker to receive job offers from corporate or governmental security agencies or for a security company to become involved in actions of questionable nature (especially when considering the difficulty of evaluating the action of a web security company which may operate in various countries according to a singular ethical and legal framework). Acts that may seem legally or morally dubious are acceptable within hacker culture, where action is often judged not in juxtaposition with (moral) intent but by merit of skill and success alone.

This distinction is important when we later consider members of the Anonymous collective, who base their identity around the content they produce and consume or around their political mobilization rather than around proficiency with technology. One does not simply "switch sides" when being part of the Anonymous collective, and unlike hacker culture, Anonymous does not allow the same freedom of individual choice regarding goals and their execution.[4]

We Are Legion: Anonymous in the Making

It is difficult to begin and describe how Anonymous came to be no less than to try and pinpoint what it is. It appears to have originated from several highly idiosyncratic Web forums, IRC (Internet Relay Chat) channels, and websites dedicated to Web culture. The online venues associated with Anonymous are primarily the http://www.4chan.org (4chan) image board forum (especially, the "/b/—random" section of it) and Encyclopedia Dramatica, the online culture antithesis to Wikipedia, a wiki devoted to Internet memes, provocative language, and shocking images (Elliott, 2008). Anonymous's ethos defies definitions such as "group" or "organization." It claims to lack formal organization or leadership.[5] A press release sent on behalf (at least supposedly) of Anonymous stated the following:

> Anonymous is not a group, but rather an Internet gathering. Both Anonymous and the media that is covering it are aware of the perceived dissent between individuals in the gathering. This does not, however, mean that the command structure of Anonymous is failing for a simple reason: Anonymous has a very loose and decentralized command structure that operates on ideas rather than directives. (ANONYMOUS, 2010, p. 1)

Anonymous's origins can be tracked down in the days before Web 2.0 neo-liberalist culture (Jarrett, 2008; O'Reilly, 2005), when the Web was still inhabited by mostly disembodied and nameless entities. Before Google suggested to individuals to open e-mail under their own names, before Amazon connected one's shopping habits to his or her credit card account, and before Facebook forbade the use of fake names in the creation of profiles, the code of conduct of the Web perceived anonymity as the norm. Human-computer interaction researcher Jacob van Kokswijk (2008) stated that online anonymity is not a new thing but a carryover of a long tradition of masks and pseudonyms, which hails from sources as varied as the Greek drama or the Renaissance carnivals. He followed Simon Biggs (1997) in observing that anonymity is often tied with attempts to examine the relations among social, personal, and bodily potential, especially as we transition through various life stages. According to van Kokswijk, Since the early days of the Internet, the value of online anonymity has been held in high regard, but it was often conceptualized more as pseudonymity, tied to the exploratory process of becoming, via maintained multiple identities. Recent new requirements in online identity management by the large corporations, coupled with legislation attempts to make online conduct more traceable (discussed in the last part of this chapter), make many uncomfortable with the rising levels of visibility. Anonymous rhetoric often invokes this notion of anonymity, especially since much of its activity occurs on platforms predating Web 2.0.

However, and despite the claims made by the collective in attempts to link this "classic" type of web anonymity with its modern values of fighting censorship and ensuring online freedoms, the anonymity of Anonymous is not akin to the classic (pseudonymity) definition, when it meant the ability to maintain a consistent alias or persona without having to identify your credentials in physical space. Rather, Anonymous promotes a vision of itself as a disembodied collective voice, in which the actions and considerations of the individual are meaningless (and, following that logic, no one is truly accountable or represents the collective as a whole). This becomes clearer when looking at the posting and discussion methods of 4chan, the primordial hatching pool of Anonymous culture. Blogger and social media critic Jana Herwig (2011) noted the following in her account of 4chan:[6]

> While conventional anonymity online meant that one's real name and identity were protected through the use of (unique and/or registered) nicknames, 4chan takes this one step further: Because no one can register, no one may claim a nickname for him or herself. (p. 11)

This is an important observation. Anonymous is not rooted in ideology of anonymity *per se* but rather in one which promotes lack of identity. On 4chan, unlike social networking sites or forums, there are no permanent identities (true or otherwise), no "social graphs" for friends/followers, or "feeds" of content connected to your persona. Each post on the image board has a unique identifier, but this 9-digit code is the only reference possible on the website. Posts are deleted after a period of hours or days (depending on popularity) and no archival record remains on the site. Although users posting or replying to a thread might assume an alias in the process of composing, this is not required and is in fact discouraged. The effect is endless boards of images and text, coming from predominantly "Anonymous" (non-named) posters. As Galloway's (2004) work suggested, this is an example of protocol shaping social interaction. Without a means to distinguish among the users on the board, the associated feeling is of a huge hive-mind communicating with itself, a single, yet heterogeneous organism, or perhaps a schizophrenic arguing with multiple personalities.

This notion is indicated in Anonymous's unofficial motto: "We are legion." Taken from the New Testament, this quote references the submergences of multiple identities in one entity, existing as a whole but disappearing when trying to pinpoint individuals. Unlike the hacker culture previously discussed, Anonymous culture seeks no recognition, intellectual or otherwise.

> Anonymous is not a person, nor is it a group, movement or cause: Anonymous is a collective of people with too much time on their hands, a commune of human thought and useless imagery. A gathering of sheep and fools, assholes and trolls, and normal everyday netizens. An anonymous collective, left to its own devices, quickly builds its own society out of rage and hate. . . . As individuals, they can be intelligent, rational, emotional and empathetic. As a mass, a group, they are devoid of humanity and mercy. Never before in the history of humanity has there once been such a morass, a terrible network of the peer-pressure that forces people to become one, become evil. Welcome to the soulless mass of blunt immorality known only as the Internet (Encyclopedia Dramatica, n.d.)

Another aspect of Anonymous culture is its apparent nihilism. The main reason for the collective to set into action is "lulz" (Bair, 2008)—the continuous search for entertainment through the pursuit of the awkward, bizarre, and unconventional, often at the expense of others. A corrupted version of the infamous web abbreviation for "Laughing Out Loud" (LOL), lulz is the reason for invading en mass another forum for relentless spam comments ("trolling") or for launching a worldwide protest against Scientology (Coleman, 2011; Elliott, 2008; Schultz, 2008). The goal may be righteous or not, the targets may "commit crimes" against Anonymous (or "the Internet" as a whole) or just be on the wrong server at the wrong time—if it provides entertainment, it is worth doing. Gabriella Coleman (2010) suggested that this dualistic behavior is in line with the cultural archetype of the trickster deity—an agent of chaos, capable of both acts of kindness and cruelty for its own amusement.

The last moment in Anonymous culture worth exploring is the way meanings are generated through the inception of memes. A meme (Dawkins, 1976) is the smallest unit of cultural information; it can be an idea, a fashion, or an architectural style. Online, the word came to signify a joke, a phenomenon, or a catchphrase. Anonymous thrives on memes, and 4chan is considered to be a central "meme-factory" for the rest of the Web (Herwig, 2011). One such example is "lolcats" (pasting misspelled comments on top of animal pictures), which originated on the boards and became a recognized Internet phenomenon.

What's interesting about memes, especially in the early stages of their origination and insemination, is the fact that they truly evolve via natural selection. There is no democratic process, voting mechanisms, or leaders who say what a meme is and what it is not. In fact, since 4chan lacks archival memory and the only way to preserve content from the website is by copying and pasting it and saving it onto individual hard drives, memes are prone to oblivion. Only by crossing a certain invisible threshold of acceptance does a meme continue its existence. It is a unique process, which may reflect something of Jodi Dean's (2003, p. 108) envisioned "neodemocracies" of the future political Web, where consensus is achieved through struggle and contestation among opposing views.

To sum up, Anonymous is not only anonymous, but also in many ways, identity-less. Its ethos is somewhat nihilistic, including self-derogatory rhetoric and action based on a fun factor rather than specific values or agendas (except, perhaps, the value of online anonymity). And last, ideas, agendas, and motivations compete within the collective for dominance, and when one "infects" the critical mass of brains required, the collective as a whole acts upon it. This factor is crucial when discussing Anonymous as political mobilization.

Politically Active

The first time Anonymous climbed from the (relative) obscurity of the Internet and into the headlines was due to its involvement with the Church of Scientology. Enraged by the church/sect's attempt to remove a leaked internal video, and interpreting it as a violation of "the laws of the Internet," the Anonymous movement decided to fight back. The struggle included both online hacktivism against Scientology's websites and off-line demonstrations in which small masked groups of Anonymous members disrupted Scientologists in hundreds of locations around the globe[7] (Bair, 2008; Elliott, 2008; "The Following Post Is [about] Anonymous," 2008). The name that emerged for this anti-Scientology campaign was "Project Chanology," a portmanteau of "scientology" and "4chan."

In the case of Project Chanology, one should note how the authorities treated those outbursts against Scientologists: they mostly ignored them. Several demonstrations were dispersed, but generally, local and state authorities declined to intervene in what appeared to be a conflict between two subcultures (Arnoldy, 2008). Furthermore,

Scientologists' "fair game" approach (Urban, 2006) openly declared the intent to persecute those who oppose it via physical, juridical, and public relations means. Anonymous's masked activists protest left Scientologists without real names or faces to target. Project Chanology, to take it from a previously discussed perspective, was very "Hackerish": a creative, even if somewhat rogue, solution to a problem that cannot be tackled by other means and which comes to pass in a peripheral field unattached to mainstream politics. To put it bluntly: both Scientology and Anonymous were too far away from the interests of the powers-that-be for them to pay attention.

This was not the case with "Operation Payback." It began as an anticopyright campaign targeted against organizations persecuting pro-piracy activists, but it was then retargeted as an online "artillery support" for Julian Assange and the WikiLeaks organization (Correll, 2010). WikiLeaks, which released hundreds of thousands of classified U.S. documents online in the preceding month, was being treated as a criminal organization by the United States and several European states. Following that, several large companies such as Amazon and Visa withdrew their dealings with WikiLeaks and froze their accounts. Anonymous responded by a call tallying its supporters to "avenge Assange,"[8] by actively propagating WikiLeaks' cause and by participating in Distributed Denial of Service (DDoS)[9] attacks on the offending companies. The attacks were carried out by a web- and software-based tool named LOIC, which allowed anyone to join in the assault without previous technical skills (Pras et al., 2010).

Later on, Anonymous hacked the accounts of Internet security firm HBGary and produced compromising e-mails in which the company supposedly planned cyber and smear attacks against WikiLeaks (Cook & Chen, 2011). In contrast to its earlier operations, Anonymous targeted rather mainstream organizations and corporations. This incident can be seen as the beginning of the "hacking-as-leaking" phase of Anonymous political activism (Coleman & Ralph, 2011), in which hacks began to commence for purposes of disclosing information, often perpetrated against institutions such as major financial and governmental agencies. From this point, Anonymous attracted a growing number of followers, with some arguably being more interested in its mantle as "hacktivists" rather than its traditional role as "provider of LOLs."

Anonymous's distributed, meme-based, decision-making process proved to be effective in dealing with real-time current affairs situations. New technology allowed everyone who wished to participate in cyber attacks to do so, eliminating the previously needed expertise-based "hacker" mantle. The reaction of authorities this time was strikingly different. Several activists were tracked and arrested (Cook & Chen, 2011), and the U.S. Federal Bureau of Investigation began an investigation of the attacks (Sandoval, 2010).

Governmental Responses

I would like to claim that the main reason for a change in the behavior of governments with regard to Anonymous lies in a larger change of the Web's role in modern society (Chadwick, 2008). The previously discussed tools and methods of organization are opening up technological activism to those previously incapable of doing so, and as society becomes more reliant on ICT, this form of activism appears increasingly problematic, as it is intertwined with the very foundations of the new economy.

Web-based giants try to assault online anonymity to achieve better segmentation of their users and turn profits. To do so, they seek governmental support "in order to preserve their property rights in the internet-based economy" (Castells, 2001, p. 181). This economy demands active participation of users in content generating and sharing platforms of Web 2.0 (Scholz, 2008). Anonymous and its activists demonstrate that this online participatory power can be used for disruption as well as for consumption.

Recent legislation shows that both the United States and the European Union have begun to take cyberspace quite seriously. Laws are being drafted to regulate cyber security, cyber crime, and one's management of identity online. Over 50 pieces of legislation have been discussed by the U.S. Congress alone over the last 2 years (Hathaway, 2010). Recent legislation includes acts like the Cybersecurity and Internet Freedom Act (Sen. Lieberman, 2011), which regulates the responsibility for ICT crimes and attacks in the United States and proposes, among other things, a "kill switch" for the president that allows cutting access to the Net infrastructure on a massive scale. Another example is the Cyber Security and American Cyber Competitiveness Act (Sen. Reid, 2011), which also deals with the issues of privacy and regulates economic entities (albeit guarantying that no restrictions would be put on the ability of federal bureaus in accessing this information). This process has reached culmination with the SOPA-Stop Online Privacy Act (Rep. Smith, 2011), which threatens to put the control over websites suspected of enabling illegal content into the hands of law enforcement agencies, to the point of preventing online service providers, such as payment services like PayPal or e-retailers like Amazon, from working with them, and search engines from linking to them, effectively removing them from the active Web.

Outside the United States, the situation is no different. The European Union has commenced the European Network and Information Security (EU, 2004), which has been ratified slowly but surely in member countries, giving more and more power to governments over cyberspace. Russia is taking control of the digital world even more seriously. A recent example is a statement by the Russian head of the FSS (the successor of the notorious KGB), in which he suggested forbidding, inside Russia, services that use internal encryption, such as Gmail, Hotmail, and Skype (Faulconbridge, 2011).[10]

Conclusion

It would be far too presumptuous to suggest that Anonymous alone is responsible for this trend in governmental legislation. But it is a very visible instance of a larger problem for the powers that be. As the Web becomes further entangled with everyday practices of commerce and politics on the one hand, and average citizens gain disruptive collective powers and technology (such as the LOIC DDoS tool) on the other, governments rightfully fear loss of control from the hands of various smart mobs. Their response is to try and return this control to themselves, for limiting it on others.

Unlike the "Hacker Culture" epoch, when the quirks of a handful of skilled mavericks could be tolerated in exchange for the potential benefits of their talents, Anonymous is not comprised of hackers. Although some of its members, without a doubt, display above average skills in computer and network systems, they do not seek personal recognition, and their end goal is (anti)social upheaval rather than improvement of existing technology. As a political entity Anonymous is, as it claims, a decentralized hive-mind structure. It is neither a group, nor an organization, nor a cause. A self-proclaimed "gathering" is indeed a fitting name. A better name, in van Dijk's (2005) terms, might simply be "a network." Or, in ANT terms, it just refuses punctualization.

Anonymous may see itself as operating for the greater good of the average user. But on the playground of the modern Web, its actions serve to emphasize how ordinary netizens have become the problem that hackers never were. The efforts perpetuated by Anonymous, especially under the chaotic, countercultural shroud it exhibits today, may in fact lead to the strengthening of online governmental control rather than helping the struggle to resist it.

Notes

1. Latour suggests substituting "networks" for "work-nets" in order to prevent confusion with terminology from Information-Communication Technologies fields such as the Internet. "Work-nets" also emphasizes the diffusion of agency (work) along the chain (net) of relations.
2. In one instance, per a journalist's request to gain access to "inner circles" of the group, members discuss amongst themselves the possibility of creating a fake Internet Relay Chat (IRC) channel with bogus code words and displaying it to her.
3. There are several examples of how the members of Anonymous do not consider themselves

hackers, although some individuals and perhaps even leaders (as much as the term applies) within this collective exhibit hacker characteristics. One is a press release (ANONYMOUS, 2010) originated from the group which states, "Anonymous is not a group of hackers. We are average Internet Citizens ourselves and our motivation is a collective sense of being fed up with all the minor and major injustices we witness every day." Another example comes from Encyclopedia Dramatica, one of the chief online collaborative forums associated with the group: "Anonymous can be anyone from well-meaning college kids with highly idiosyncratic senses of humor trying to save people from Scientology, to devious nihilist hackers, to clever nerds, to thirteen year old boys who speak entirely in in-jokes on an endless quest for porn . . . ."

4. In contrast to what may seem, actions by individuals that deviate from the Anonymous ethos (discussed later) or call for such actions are frowned upon and may lead to a collective backlash against the individual. This is also known as "not your personal army" web meme (Encyclopedia Dramatica, n.d., b).
5. Leaked protocols of online meetings have later shown that this is not completely true, and some leadership is allocated (or assumed) for specific tasks. I will deliberate on this in the next part.
6. Despite its idiosyncratic content, it attracts about 9.5 million unique users monthly (Herwig, 2011).
7. Anonymous *modus operandi* was in organizing quick, distributed, cell-based flash attacks, best described by Howard Rheingold's concept of *smart mobs* (Rheingold, 2002).
8. The complete text of the message can be seen here: http://en.wikipedia.org/wiki/File:Avenge_Assange_Anonymous.png
9. Distributed Denial of Service attacks are simple methods for disrupting access to a specific website by bombarding it in server requests sent from multiple computers, thus leaving the server unable to deal with incoming traffic.
10. This chapter's scope is far too narrow to include all possible examples on cyber legislation from recent years, and just the discussion of steps taken in more authoritative regimes (China in particular) may constitute a paper of its own. These examples are meant to show the increasingly active role governments try to assume in cyberspace as well as their attempt to expropriate the control over online life from private users and commercial entities.

References

ANONYMOUS. (2010, October 10). ANON OPS: A press release. Retrieved from http://www.wired.com/images_blogs/threatlevel/2010/12/ANONOPS_The_Press_Release.pdf

Arnoldy, B. (2008, March 17). Anonymous activists gaining strength online. CSMonitor.com. Retrieved from http://www.csmonitor.com/USA/Society/2008/0317/p03s02-ussc.html

Bair, A. (2008). "We are legion": An anthropological perspective on Anonymous. *Proceedings of the 2008 Senior Symposium in Anthropology*, Department of Anthropology, Idaho State University. 41–48. Pocatello: Idaho State University.

Barabási, A.-L. (2002). *Linked: The new science of networks*. Cambridge, MA: Perseus Pub.

Biggs, S. (1997). Choosing not to be old? Masks, bodies and identity management in later life. *Ageing & Society*, *17*(05), 553–570. doi:null

Castells, M. (2001). *The internet galaxy: Reflections on the internet, business and society*. New York, NY: Oxford University Press.

Chadwick, A. (2008). Web 2.0: New challenges for the study of e-democracy in era of informational exuberance. *I/S: Journal of Law and Policy for the Information Society*, *5*, 9.

Coleman, G. (2010). Hacker and troller as trickster. Retrieved from http://www.socialtextjournal.org/blog_dev/2010/02/hacker-and-troller-as-trickster.php

Coleman, G. (2011, April 6). Anonymous: From the lulz to collective action. The New Everyday. Retrieved from http://mediacommons.futureofthebook.org/tne/pieces/anonymous-lulz-collective-action

Coleman, G., & Golub, A. (2008). Hacker practice. *Anthropological Theory*, *8*(3), 255–277. doi:10.1177/1463499608093814

Coleman, G., & Ralph, M. (2011). Is it a crime? The transgressive politics of hacking in Anonymous. Social Text Journal Blog. Retrieved from http://www.socialtextjournal.org/blog/2011/09/is-it-a-crime-the-transgressive-politics-of-hacking-in-anonymous.php

Cook, J., & Chen, A. (2011, March 18). Inside Anonymous' secret war room. Gawker. Retrieved from http://gawker.com/#!5783173/inside-anonymous-secret-war-room

Correll, S.-P. (2010, December 6). Operation: Payback broadens to "Operation Avenge Assange." Panda Security. Retrieved from http://pandalabs.pandasecurity.com/operationpayback-broadens-to-operation-avenge-assange

Dawkins, R. (1976). *The selfish gene*. New York. NY: Oxford University Press.

Dean, J. (2003). Why the net is not a public sphere. *Constellations*, *10*(1), 95–112. doi:10.1111/1467–8675.00315

Elliott, D. C. (2008). Anonymous rising. *LiNQ*, *36*, 96–111.

Encyclopedia Dramatica. (n.d. -a).Anonymous. Encyclopedia Dramatica. Retrieved from http://encyclopediadramatica.se/Anonymous

Encyclopedia Dramatica. (n.d. -b). X_is_not_your_personal_army. Encyclopedia Dramatica. Retrieved from http://encyclopediadramatica.se/X_is_not_your_personal_army

EU. (2004). Regulation (EC) No 460/2004 of the European Parliament and of the Council of 10 March 2004 establishing the European Network and Information Security Agency. 460/2004. Retrieved from http://eur-lex.europa.eu/LexUriServ/LexUriServ.do?uri=CELEX:32004R0460:EN:NOT

Faulconbridge, G. (2011, April 8). Russian spy agency complains about Gmail, Skype. Reuters. Retrieved from http://www.reuters.com/article/2011/04/08/us-russia-internet-idUSTRE7374C720110408

Galloway, A. R. (2004). *Protocol: How control exists after decentralization.* Cambridge, MA: MIT Press.

Hathaway, M. (2010, November). Cybersecurity: The U.S. legislative agenda part II. Retrieved from http://belfercenter.ksg.harvard.edu/files/short-summary-legislation-nov2010.pdf

Herwig, J. (2011). The archive as the repertoire mediated and embodied practice on imageboard 4chan. org. Retrieved from http://homepage.univie.ac.at/jana.herwig/PDF/Herwig_Jana_4chan_Archive_Repertoire_2011.pdf

Jarrett, K. (2008). Interactivity is evil! A critical investigation of Web 2.0. *First Monday*, *13*(3), 34–41.

Kokswijk, J. V. (2008). *Digital ego: Social and legal aspects of virtual identity.* Delft, South Holland, The Netherlands: Eburon Publishers.

Lanier, J. (n.d.). Brief biography of Jaron Lanier. Retrieved from http://www.jaronlanier.com/general.html

Latour, B. (1987). *Science in action.* Cambridge, MA: Harvard University Press.

Latour, B. (2005). *Reassembling the social: An introduction to actor-network-theory.* Oxford, UK: Oxford University Press.

Law, J. (1992). Notes on the theory of the actor-network: Ordering, strategy, and heterogeneity. *Systems Practice, 5*(4), 379–393. doi:10.1007/BF01059830

Levy, S. (1984). *Hackers: Heroes of the computer revolution.* Garden City, NY: Anchor Press/ Doubleday. Retrieved from http://lib.agu.edu.vn:8180/collection/handle/123456789 /2644

O'Reilly, T. (2005). What is Web 2.0: Design patterns and business models for the next generation of software. Retrieved from http://www.oreillynet.com/pub/a/oreilly/tim/news/2005/ 09/30/what-is-web-20.html

Pras, A., Sperotto, A., Moura, G. C. M., Drago, I., Barbosa, R., Sadre, R., et al. (2010). Attacks by "Anonymous" WikiLeaks proponents not anonymous. Design and Analysis of Communication Systems Group (DACS), University of Twente, Enschede, The Netherlands. Retrieved from http://eprints.eemcs.utwente.nl/19151/01/2010–12-CTIT-TR.pdf

Rep. Smith, L. (2011). H.R.3261—Stop online piracy act. H.R.3261. Retrieved from http://thomas.loc.gov/home/gpoxmlc112/h3261_ih.xml

Rheingold, H. (2000). *The virtual community: Homesteading on the electronic frontier.* Cambridge, MA: MIT Press.

Rheingold, H. (2002). *Smart mobs: The next social revolution.* Cambridge, MA: Perseus Pub.

Roversi, A. (2008). *Hate on the net: Extremist sites, neo-fascism on-line, electronic jihad.* Bodmin, UK: Ashgate Publishing.

Sandoval, G. (2010, November 9). FBI probes 4chan's "Anonymous" DDoS attacks. CNET News. Retrieved from http://news.cnet.com/8301–31001_3–20022264–261.html

Scholz, T. (2008). Market ideology and the myths of Web 2.0. *First Monday, 13*(3), 3.

Schultz, D. (2008, February 15). MediaShift idea lab. Anonymous vs. Scientology: A case study of digital media. Retrieved from http://www.pbs.org/idealab/2008/02/anonymous-vs-scientology-a-case-study-of-digital-media005.html

Sen. Lieberman, J. I. (2011). S.413—Cybersecurity and internet freedom act of 2011. Retrieved from http://thomas.loc.gov/cgi-bin/bdquery/z?d112:S.413.

Sen. Reid, H. (2011). S.21.IS—Cyber security and American cyber competitiveness Act of 2011. Retrieved from http://thomas.loc.gov/cgi-bin/query/z?c112:S.21.

The Following Post Is [about] Anonymous. (2008, April 3).Confessions of an Aca/Fan. Retrieved from http://henryjenkins.org/2008/04/anon.html

Urban, H. B. (2006). Fair game: Secrecy, security, and the Church of Scientology in cold war America. *Journal of the American Academy of Religion*, 74(2), 356–389. doi:10.1093/ jaarel/lfj084

van Dijk, J. (2005). *The network society: Social aspects of new media* (2nd ed.). London, UK: Sage.

Wayner, P. (2000). Free for all: How Linux and the free software movement undercut the high-tech titans. Harper Business. Retrieved from http://www.jus.uio.no/sisu/free_for_all.peter _wayner/landscape.a5.pdf

SECTION 4

Emerging Issues in Digital Ethics

INTRODUCTION BY BASTIAAN VANACKER

"Is evil (or good) something you are or something you do?" is a question that has occupied philosophers for ages but has been given a new shine of relevance through the rise of modern technology and the digital (r)evolution.

Technological advancements, for example, have made us question the nature of evil. Who is responsible when a drone kills innocent civilians in Afghanistan? Who is to blame when a news aggregator mistakenly selects an old headline about an airline's near-bankruptcy and presents it as current news, causing its stock to plummet? How can our ethical systems relying on deontology, social contract theory, or the principle of utility begin to evaluate these "new" instances of evil, which seem to defy traditional distinctions between natural and moral evil? Floridi and Sanders (2001) claimed that they cannot, and they proposed the development of an Information Ethics (IE) as a sui generis ethical theory to deal with these new types of ethical problems.

Anthony Beavers places his article in this tradition. Motivated by a belief that a "no-fault ethics" is needed to confront the new problems we are facing, he challenges some of meta-ethics' most deeply held beliefs. Beavers argues that moral interiority–the notion that a true moral agent not merely does the right thing,but does it consciously, out of free will and intentionality—is not necessary for ethics. Using the terra firma provided by computational ethics, he argues that since moral agency requires that a moral agent be responsible and accountable for his/her/its actions, it also must require that this moral agent be capable of both succeeding and failing in his/her/its moral obligations. But because we have a moral obligation not to unleash upon the world machines capable of failing in their moral obligations, logic mandates that we cannot want machines to be responsible and accountable, that is, having moral agency (or interiority).

The practical implications of this last premise (that we have a moral obligation not to manufacture machines capable of moral failure) might present a challenge to engineers, as producing infallibility—moral or other—is a seemingly impossible standard to meet for mere mortals. Practical objections aside, his impenetrable logic leads Beavers to a fascinating, if slightly unsettling, conclusion: If we could design an ethical decision-making machine (MorMach) that possesses perfect ethical decision-making calculation skills, it would be our moral obligation to show deference to it. Doing the right thing could be reduced to mere mindless following of MorMach's instructions. This seems counterintuitive; mindless rule following has contributed to some of the darkest pages of the 20th century. But it is particularly its ability to challenge our intuitions through logical reasoning that makes this contribution such a rewarding piece of philosophy.

In addition to technological advancements mentioned earlier, digital worlds and computer games also have brought to the forefront the question about the nature of good and evil by challenging our notions of identity, moral agency, and of ethics in general. Julian Dibbell's seminal 1993 article, "A Rape in Cyberspace," for example, describes the hurt and outrage caused by one rogue member of a Lambda MOO (a text-based virtual community) group who engaged in virtual sexual abuse of fellow MOOers. The article, which first appeared in *The Village Voice,* has reached a canonical status among Internet researchers. Dibbell's luncheon presentation on the legacy of this very article at the conference that served as the basis for this volume speaks to its ongoing relevance. While by no means a dystopian account, Dibbell's article did draw attention to the darker side of the Internet and virtual worlds. Others, such as Jane McGonigal, the keynote speaker at that same conference, have stressed the beneficial effects of gaming.

It is against these latter utopian visions of gaming and gamification that Miguel Sicart revolts in his chapter. While convincingly exposing the weaknesses of the gamification argument—a misinterpreting of the ludic elements of play and too great an indebtedness to scientificism and instrumentality—Sicart resists the temptation to give in to a dystopian account of games as ethical wastelands. Instead, he argues for design choices that fulfill the needs of the "homo poieticus," a typology borrowed from Floridi and Sanders's (2001) IE, replacing the "homo ludens." The homo poieticus is a player who is no longer passive. He or she considers the creativity involved within play not a by-product of the activity but a requirement. In fact, the homo poieticus, Sicart argues, "places ethical value in the act of the creative appropriation."

The practical applications of this theoretical insight lay in the challenge it presents for designers. For Sicart, designers cannot be satisfied with the notion that (ethical) skills acquired in game play can then be applied to other domains than games. He challenges designers and philosophers to come up with technologies that make play practical ethical development rather than a practice course for the real world, as the instrumentalist account of gamification stipulates. How this can be done is as of yet an open question, but the interplay among design theory, game theory, and an IE-based anthropology explored here by the author provide an interesting pathway to put the "play" back into gamification.

References

Dibbell, J. (1993). A rape in cyberspace. Retrieved from http://www.juliandibbell.com/articles/a-rape-in-cyberspace/

Floridi, L., & Sanders, J. (2001). Artificial evil and the foundations of computer ethics. *Ethics and Information Technology*, *3*, 55–66.

CHAPTER TWELVE

Could and Should the Ought Disappear from Ethics?

ANTHONY F. BEAVERS

A Provocation

In his 1961 monograph, *Totality and Infinity: An Essay on Exteriority*, the late phenomenologist, Emmanuel Levinas (1961/1969), noted that, "everyone will readily agree that it is of the highest importance to know whether we are not duped by morality" (p. 21). What follows thereafter is an extensive attempt to ground a quasi-Kantian existential ethics based on interpersonal, face-to-face relations (Beavers, 2001). That philosophy should invite such an attempt already signifies that we might be in trouble where ethics are concerned. After all, when one looks back over history, it appears that we have not made much progress in this area. We still fight senseless wars, accepting them passively as an inevitability of the human condition; and as we move more deeply into the information age, any emerging prospects for peace seem to be losing ground due to new mechanisms of power and social control brought about in part by the increasing dehumanization of the human, as "persons" become interchangeable with "profiles," and in part, by the ubiquity of information, new surveillance technologies, long-distance robotic weapons, and "cubicle warriors" (see Royakkers & van Est, 2010). Indeed, the politics and economics of information and information flow are quickly becoming the intermediaries of our social connections. Even as early as 1982, the same year that *Time Magazine* named the personal computer *person* of the year and 12 years before the

World Wide Web, Levinas could see things changing:

> The problem that concerns us in this conference—that of the community—is, without doubt, a topical one, due to the unease felt by man today within a society whose boundaries have become, in a sense, planetary: a society in which, due to the ease of modern communications and transport, and the worldwide scale of its industrial economy, each person feels simultaneously that he is related to humanity as a whole, and equally that he is alone and lost. With each radio broadcast and each day's papers one may well feel caught up in the most distant events, and connected to mankind everywhere; but one also understands that one's personal destiny, freedom or happiness is subject to causes which operate with inhuman force. One understands that the very progress of technology—and here I am taking up a commonplace—which relates everyone in the world to everyone else, is inseparable from a necessity which leaves all men anonymous. Impersonal forms of relation come to replace the more direct forms, the "short connections" as Ricoeur calls them, in an excessively programmed world. (1982/1989, p. 212)

Of course, these observations came from a day when mass media were largely a one-way affair. Several of us hoped in 1994 that the appearance of the World Wide Web, that is, *two-way* media, might replace the pessimism that Levinas announced here, that the world in which humans "find themselves side by side rather than face to face" (1982/1989, p. 212) might be redeemed as we learned to confront each other as individuals once again. Such has not been the case. If anything, the sheer explosion of easily accessible information has us lost even more as our objective presence in the form of "profiles" and "voices" supplants the subjective interiority that formerly motivated our ethical concerns and anesthetizes us to the real impact of our actions that hide behind representations that casually take their place. Furthermore, nothing stands still long enough to consider in detail, in any case, and even if it did, what difference could one voice make or a million in a social network with 750 million members? After all, it is the aggregate that counts. Is this some type of emergent democracy in action or a genuine threat to the fabric of our moral lives?

Though somewhat rearranged, this question is not new. Signs of deep ethical change were already apparent in several philosophers of the 19th century, particularly in Nietzsche and Kierkegaard, who both announced the emergence of the objective human and mourned the loss of the subjective individual. *Truth* was their culprit; *information* is ours. Either way, knowledge or information about me, my public profile, is what must be managed, while the development of my moral character behind it goes unnoticed or, worse yet, is flat-out irrelevant. Anymore, it's all about the appearances, and one small slip can ruin a career (or, with luck, make you more famous, as with Bill Clinton). Given that appearances are all we can see anyway, we can and should ask whether this is good or bad from a moral point of view.

A pessimistic answer has been painted thus far, but if ethics, traditionally understood, has not served us well, perhaps a change of perspective might be in order.

Though it claims to be an "essay on exteriority," Levinas's (1961/1969) *Totality and Infinity* is really an essay on interiority, on one's moral subjectivity. This focus seems appropriate for one trying to determine whether we have been duped by morality, and it will be mine here too, as I address the rather daring question, "*Could* and *should* the *ought* disappear from ethics?" The words in italics are all past subjunctives, which in English means that they refer to unreal conditions; they refer to what could be, should be, or ought to be, not about what is. It is easy to get lost in flights of fancy when addressing the unreal. So, to address the broader question of whether we have been duped by morality, we need to get our feet on the ground. That said, it's time to leave the continent in search of solid ground.

Terra Firma

In 2007, Anderson and Anderson wrote, "As Daniel Dennett (2006) recently stated, AI 'makes philosophy honest.' Ethics must be made computable in order to make it clear exactly how agents ought to behave in ethical dilemmas" (p. 16). To rephrase their sentiment, a computable system or theory of ethics serves to make ethics honest. As I have observed elsewhere (Beavers, 2010), it is common among machine ethicists to note that research in computational ethics can help us better understand ethics in the case of human beings. This is because of what we must know about ethics in general to build machines that operate within normative parameters. Unclear intuitions will not do where engineering specifications and computational clarity are required. So, machine ethicists are forced head on to engage in moral philosophy. Their effort, of course, hangs on a careful analysis of ethical theories, the role of affect in making moral decisions, relationships between agents and patients, and so forth. But this is not all. There are other meta-ethical difficulties that must be addressed as well concerning, particularly, the nature of the moral ought and the necessary and sufficient conditions for moral agency. Every moral theory makes assumptions about these issues, but, to date, without the clarity that real-world, working specifications for practical application require. Thus, computational ethics provides us with the *terra firma* needed to get some solid footing in the otherwise vague and messy domain of ethics and helps us answer the question of whether (and if so, to what extent) we may have been duped by morality.

My conclusion here will be that yes, we have been duped, at least in part. As such, I am departing from Beavers (2009, 2011), where I suggested that we *might* (another past subjunctive) have been in order to argue for something more definite. This conclusion will be an unhappy one for many, since it will involve throwing out an age-old distinction between being good and merely acting so that has been at the heart of ethics (according to the dominant Western paradigm) from its inception. My argument will unfold in three parts: the first will address the question of moral agents (MAs) in general, after which I will examine what precisely *ought* implies when viewed from a moral perspective to isolate what I will identify as the paradox of automated moral agency (P-AMA). Next, to avoid the paradox, we will need to define the *ought* technically, not morally, a distinction I am partially borrowing from Kant and will make clear later, with the result that we are left with the sufficiency argument (SA), which states that moral interiority is a sufficient but not necessary condition for moral agency. If this argument holds, then the kind of moral interiority that allows an agent to be culpable for its actions is not necessary for ethics. It is rather a product of our biology that drives humans to be ethical, but it is not the only way this can be done. My motive in taking this approach is that the problems we are facing as a world are so great that it is time to put aside the "blame game" and confront them with some sort of no-fault ethics, the details of which have largely been worked out by Floridi (1999, 2002) and Floridi and Sanders (2001, 2004), though more work needs to be done here to draw out the implications where blame and fault are concerned.

Moral Agency

A taxonomy of moral agency has already been worked out by Moor (2006) and consists of (a) ethical-impact agents, (b) implicit moral agents, (c) explicit moral agents, and (d) full moral agents, ranked in increasing order as they approach genuine moral agency. Almost any machine can be an ethical-impact agent, since all that is meant here is that the machine may have a straightforward moral impact. Since this includes (almost?) all machines, we will not be concerned with ethical-impact agents in this paper.

Implicit ethical agents are machines constrained "to avoid unethical outcomes" (Moor, p. 19). They do not work out ethical solutions for themselves, but they are constrained by design to act in a morally respecting way. Moor mentions automated teller machines (ATMs) and automatic pilots as examples. The ATM isn't programmed with a rule to act fairly any more than the automatic pilot must decide to act safely to spare human life. Explicit ethical agents, on the other hand, are machines that can "'do' ethics like a computer can play chess" (Moor, pp. 19–20). They apply principles to concrete situations and decide how to act. The principles might be something like Kant's categorical imperative or Mill's principle of utility. The critical component of explicit ethical agents is that they work out ethical decisions for themselves using some kind of recognizable moral decision procedure. Such machines should therefore be able to answer for their behavior. Finally, full ethical agents are beings like us, with "consciousness, intentionality, and free will" (Moor, p. 20). They can be held accountable for their actions—in the moral sense, they can be at fault—precisely because their decisions are in some rich sense up to them.

This taxonomy reappears in several places in the literature and has almost become canonical. But it suffers from an implicit bias in that it is a ranked order with human-like agents at the top. The taxonomy clearly suggests that full, yet artificial, moral agency would be ideal for robots, and that robot ethicists should work in this direction, but without addressing the question about whether placing such machines in the world would be desirable or whether it would even be moral to do so. In other words, the taxonomy is anthropocentric. In its place, I wish to suggest something more neutral so that we can better assess the prospects and goals of moral machines.

I start here, then, with the general notion of a moral agent (MA) straightforwardly defined as the following: *any agent that does the right thing morally, however determined.*

We can then divide this category into two subdivisions, responsible moral agents (RMAs) and artificial moral agents (AMAs), where an RMA is defined as *an MA that is fully responsible and accountable for its actions* (it can decide things for itself and also has the capabilities for full moral agency as defined by Moor), and an AMA is defined as *a manufactured MA that may or may not be an RMA.*

Regardless of whether we are disposed to believe that an RMA is the only kind of genuine MA, we can now ask two questions: (a) Could an AMA be an RMA, and (b) should an AMA be an RMA? The first of these is a technical question that hangs on what we might learn about consciousness, free will, and intentionality in the future and whether future technologies will allow us to build machines with these properties. Thankfully, that matter does not concern us here. The second question, however, is of immediate and pressing concern. In the next section, we will see that the answer to it needs to be negative if we are to adopt conventional notions of responsibility that hang on some sort of moral interiority.

What Ought Implies

Though Kant (1785/1993) analyzed the ought using a three-fold division involving "rules of skill, counsels of prudence, [and] commands (laws) of morality" (p. 26), it is sufficient for our purposes to stick to his broader two-fold division between hypothetical and categorical imperatives. The first two of the three just mentioned are hypothetical in that they can be understood according to an "if . . . then . . ." structure. In this paper, we can place them under the broader category of the "technical" ought (TO), which we can sharply distinguish from the "moral" ought

(MO) that governs Kant's commands of morality. The TO emerges in a moment of decision involving a choice between inclinations: *if agent X seeks a goal state Y and Z is necessary for Y, then X ought to do Z in order to get Y.* Such an ought would still hold good, if X did not want Z but wanted Y more than Z. Otherwise Z is not something X ought to pursue, *ceteris paribus.*

The MO differs from this situation in that it does not emerge as a result of a collision between inclinations or preferences but as a result of a collision between an inclination, what I want, and what morality constrains me from pursuing: *agent X is inclined to seek a goal state Y but ought not because the moral law prohibits X from pursuing Y.*

Without going into the details of Kant's text (see Kant, 1785/1993, and/or Beavers, 2009), this distinction defines the difference between heteronomy and autonomy that serves as the foundation for Kantian ethics. More importantly, it also defines the ethical situation as understood in several Judeo-Christian traditions. Indeed, the Jewish philosopher/theologian who inaugurated this chapter, Emmanuel Levinas (1977/1990), wrote the following:

> The drama of existence is not only that existence is divided into choices between desires [TO] but that existence is also suspended between the Law that is given me and my nature [MO], which is incapable of submitting to the Law without constraint. It is not freedom which defines the human being. It is obedience which defines him. (p. 166)

This distinction between the technical and moral ought will be important in what follows, particularly because the technical ought represents a decision procedure involving prioritization that describes the situation of an AMA and the moral ought as a choice between a desire and the moral law that describes the situation of an RMA. This is partly apparent in that failure to choose correctly when confronting a technical ought is a mere mistake, but failure to respond to a moral ought is an ethical transgression. Further elucidation of the significance of this difference appears when we ask whether we should want an AMA to also be an RMA. To get to this point, we need to ask what it is that ought implies.

One implication is that *ought implies can,* a maxim that is often attributed to Kant but that does not appear in his writings, though the sentiment is clearly present. Less well known but also clearly present is the notion that *ought implies might not.* This is apparent in Section II of the *Grounding for the Metaphysics of Morals,* where Kant (1785/1993) wrote, "All imperatives are expressed by an *ought* and thereby indicate the relation of an objective law of reason to a will that is not necessarily determined by this law because of its subjective constitution" (p. 24). The critical words for the current discussion are "not necessarily determined," because if the will were so determined, the (so-called) ought would not be an ought but a *would* (yet another past subjunctive). Thus, Kant wrote the following:

> A perfectly good will would thus be quite as much subject to objective laws (of the good), but could not be conceived as thereby necessitated to act in conformity with law, inasmuch as it can of itself, according to its subjective constitution, be determined only by the representation of the good. Therefore no imperatives hold for the divine will, and in general for a holy will; the *ought* here is out of place, because the *would* is already of itself necessarily in agreement with the law. Consequently, imperatives are only formulas for expressing the relation of objective laws of willing in general to the subjective imperfections of the will of this or that rational being, e.g., the human will. (p. 24)

The long and short of these passages is that the *ought* defined along moral lines (according to Kant and several strands in the Judeo-Christian traditions) belongs to human beings that are suspended between their self-centered inclinations and their greater moral obligations. It is only on the basis of such a platform that it makes sense to praise or blame an agent for its moral successes or failings. If, indeed, meeting the requirements for an agent to be worthy of praise or blame for its actions is a constituent of moral responsibility and accountability, as seems to be the case, then such an agent must remain in a position where it *might* or *might not* perform an action required by the *ought* if the *ought* is to be moral and not merely technical. It is precisely at this point where we start to run into problems with the suggestion that AMAs should be RMAs. Given that the need to make a machine an MA in the first place stems from the fact that such machines are autonomous, that is, they are self-guided, rather than act by remote control, we run into a paradox, which I identify here as the "paradox of automated moral agency" or P-AMA.

P-AMA says the following:

1. If we are to build autonomous machines, we have a *prima facie* moral obligation to (try to) make them RMAs, that is, agents that are responsible and able to be held responsible for their actions.
2. For an RMA to be responsible and able to be held responsible for its actions, it must be capable of both succeeding and failing in its moral obligations.
3. An AMA that is also an RMA must therefore be *designed* to be capable of both succeeding and failing in its moral obligations.
4. It would be a moral failure to unleash upon the world machines that are capable of failing in their moral obligations.
5. Therefore, we have a moral obligation to build AMAs that are not also RMAs.

P-AMA might be escapable as a paradox by denying either premise 1 or 4. So, a few comments are in order, first respecting premise 4.

In order for an AMA to be a genuine RMA, it is necessary that its ability to fail in its moral obligations be left unspecified. That is, an RMA cannot be responsible and held responsible for X unless it has the ability to succeed or fail in its obligations to X. It will not do to permit our AMA some innocuous moral failing respecting Y if we wish to hold it responsible for X. Thus, the reason that premise 4 holds good is that we cannot predict the extent of moral damage an RMA might unleash on the world. In fact, the extent of moral damage that an unregulated autonomous machine might unleash on the world is not predictable either, and this situation is what makes premise 1 hold good as well, unless we outright reject the paradigm of responsible agency for AMAs.

To escape from P-AMA, for instance, it would seem that the only moral way to build an AMA is not to build it as an RMA but to design it in such a way that it is incapable of moral failure, in which case, if it does fail, this would be due to a mistake or design flaw and not an ethical transgression, which is to say that it must be built to obey technical and not moral oughts. (It also means further that we cannot build *morally* autonomous machines, since they must be required by design to obey their decision procedures, in which case they are not autonomous in any deep moral sense, and also that the responsibility for the failure of an AMA falls back on the designer, but these are issues for another paper.) So, it seems that we cannot morally want an AMA to be an RMA.

From the meta-ethical perspective, however, this conclusion raises an even larger issue. If we can't morally want AMAs to be RMAs, then how can we justify wanting human beings to be RMAs? In fact, how do we avoid drawing the conclusion that RMAs are morally deficient in general? My answer is that we cannot on either point, though necessity compels us to accept what is the case; humans are as they are. But this means further that morality traditionally conceived with all the bells and whistles of full moral agency described by Moor earlier is an ethics for broken beings (or if we wish to be truer to the religious heritage that gave us ethics in the West, an ethics for "fallen creatures").

The Sufficiency Argument

In short, the sufficiency argument says that the kind of moral interiority necessary for an agent to be an RMA is a sufficient though not necessary condition for being an MA. Therefore, moral interiority is not essential for moral agency (Beavers, 2011). To accept the argument, we must acknowledge that an MA is a genuine moral agent, whether it be an AMA, an RMA, or both. There are at least three ways to do so.

The first hangs on defining MAs on the basis of their moral success as such. If RMAs are MAs, though morally deficient because *ought implies might not,* they pale in comparison to properly programmed AMAs that, while they might not be *morally* autonomous in the anthropocentric sense described earlier, may never make a moral mistake. If being a moral agent means being autonomous in the non-moral sense of being self-guiding and yet doing the right thing morally, then we are home free, since AMAs could well be, on this criterion, just as moral, if not more so, than RMAs, though without any of the moral interiority that motivates human ethics. This clearly implies a type of moral functionalism in which the mechanisms for moral action are multiply realizable. They are biologically realized in the conscience of humans who are bound by the moral ought and technologically realized in the decision procedures of machines who are bound by the technical ought. But have we begged the question, or perhaps have those who advocate for the RMAs? More on this question in a moment.

A second way to consider the question is from a linguistic point of view. In 1950, Alan Turing (1950) noted that "at the end of the century the use of words and general educated opinion will have altered so much that one will be able to speak of machines thinking without expecting to be contradicted" (p. 442). He appears to have been right about this, at least according to Kurzweil (1990). Even though the use of present language has yet to (and may never) make sense out of suggesting that my computer is thoughtful, insightful, or wise, the situation seems to change when we consider robots instead of computers. Although we do not have sufficiently sophisticated robots at present, it is easy to see that it will make sense (someday) to say that a particular robot is thoughtful, insightful, or wise or that it is morally good, conscientious, or courageous. Why might such predicates seem meaningful in the case of robots and not computers? Because, I would submit, they are applied with behavioral criteria in mind. How do I know when a robot is courageous? When it does not cringe before power. How do I know when it is conscientious? When it responds to the rights and feelings of others rather than its (simulated?) own. How do I know when it is good morally speaking? When it responds with an acute sense of propriety in all situations. Such talk exhibits a kind of moral behaviorism that is difficult to dismiss and that is already ubiquitous in our world.

Though set to a different task, for instance, namely of espousing a view very much opposed to the one I am offering here, Connolly (2011) drew an analogy between machines and corporations, which we can extend further to other collective agencies. A corporation is very much like a machine, he noted. His point is that just as we cannot predicate moral agency to a machine, we ought not predicate it to a corporation. Nonetheless, though it might be a stretch to describe a company as "not losing its soul," it is at least meaningful to suggest that a company, university, government, and so forth, can act with conscience, meaning that it employs a set of decision procedures with checks and balances that takes into account any and all of the stakeholders affected by its actions and then acts accordingly. From the view of moral behaviorism, in other words, it does make considerable sense to talk of moral and immoral governments and regimes.

A third way to address the question of whether an MA is a genuine moral agent, while simultaneously responding to the concerns about begging the question raised earlier, is to ask about how we might determine at what point an AMA deserves to be called moral in the first place. Allen, Varner, and Zinser (2000) provided a clue by positing the notion of a Moral Turing Test (MTT), which involves comparing the behavior of AMAs with RMAs. Though they presented several varieties of the test, one is particularly pertinent to this discussion: "an . . . MTT could be structured in such a way that the 'interrogator' [i.e., the judge in the experiment] is given pairs of descriptions of actual, morally-significant actions of a human and an AMA, purged of all references that would identify the agents. If the interrogator correctly identifies the machine at a level above chance the machine has failed the test" (p. 254). Interestingly, they acknowledged a limitation in this version of the test, because an AMA might be distinguishable from the human if it actually performs "consistently *better* than a human in the same situation" (p. 255). They suggested turning tables and asking the interrogator to try to determine which agent is the less moral. For present purposes, the interesting point is that we might actually have to dumb down the moral capabilities of our AMA to pass the test, and this is telling when it comes to accepting that the generic MA is a genuine moral agent. If the test encourages us to identify the *less* moral agent, here the human being, as the *genuine* moral agent, then we've already missed the ethical boat. Indeed, Allen et al. noted that "the goal of AI researchers when constructing an AMA should not be just to construct a moral agent, but to construct an exemplary or even a perfect moral agent" (p. 255). I agree, but then . . .

Conclusions

Could and should the ought disappear from ethics and to what extent have we been duped by morality? I address each question in turn, beginning with the clarification that by "ought" here, I mean the moral ought and not the technical one.

Could the moral ought disappear from ethics? The sheer possibility that a machine might be better at determining a moral course of action places the ought in a very precarious position. Complicit with this fact, we can observe threats all around in the notion that "codes of conduct" are often now considered "ethical codes" and in the further notion that one's ability for moral conformity resides somewhere in the prefrontal cortex of the brain. Given the first, that ethical behavior is becoming synonymous with adherence to a code of conduct or law, and the second, that moral fault may be a matter of faulty neural wiring and hormone control systems, the question of how we should behave morally can no longer be determined on the basis of our frail biology and failed social systems. Thus, it is not such a far-fetched idea that we might have to trust a machine for moral guidance. (After all, we've been using books to serve this very purpose for millennia.)

Let us, then, postulate the existence of such a machine and call it MorMach (short for "moral machine"). Its sole job is to calculate a moral course of action using some decision procedure (Aristotle, Kant, Mill, Rawls, etc.) or an amalgamation of several to determine what an individual should do in any given situation. Suppose that the size and scope of MorMach's dataset is such that it sees all and knows all (or almost all) and that after several succeeding generations, MorMach becomes the sole authority, the ultimate oracle, if you will, in all matters of morals.

For an individual to be moral in this scenario, deference to MorMach would be all that is required, and for those who followed its divine command without question, the responsibility for any moral transgression would fall back on MorMach itself, assuming there could be such a case. (In fact, one could even argue here that deference to MorMach should be coerced by legal sanction or, more dramatically, that humans be designed with simply no choice in the matter.) Given this situation, the moral ought (and with it all deferential RMAs) would be removed from ethics. So, yes, it would seem theoretically possible that the moral ought could disappear from ethics.

Now, should the moral ought disappear from ethics? This question is a little easier to answer, actually. If MorMach were to exist, it would seem morally imperative that we actually defer to its divine command. After all, we are, on our own, at best morally deficient RMAs with partial vision, limited computational abilities, and conflicting emotions, whereas MorMach would be morally perfect. But if this is the case, then *we already have a moral responsibility to put our efforts into building such a machine.* We *should,* in other words, work to create the situation where the moral ought *could* disappear from ethics. If this suggestion sounds surprising, it might help to realize that this is just another way of saying that the job of ethics is to make itself obsolete.

Finally, to what extent have we been duped by morality? For all of recorded history and probably long before, we have been a people torn apart by war. The promise of ethics has been to save us from this very fate, and the judgment of history must be that it has failed dramatically. A conception of the self as an RMA who, in its free will, might or might not succeed in fulfilling its moral obligations and who is in constant need of a redeemer has been an intrinsic part of this failing promise.

We've been waiting quite some time for deliverance. How much longer before we realize that the pieces of this moral puzzle just don't quite fit together? That moral behavior ought to be regulated by human biology and our limited intellects has so far been required by the necessities of history, but as the technology changes and moral machines become possible, might this appear as the ultimate deceit?

Most of the world's population is still looking to invisible and silent gods for instructions on what to do, but MorMach may be our best hope yet, and it will speak in a language that we can clearly understand.

A Palinode

I once knew a seminarian who, after spending 2 years working in a soup kitchen, decided that he should quit his work because his motives were not pure. "How selfish of me," he asked, "to use people to satisfy my need for moral cleanliness?" How selfish of him, we might ask, to put his moral purity ahead of others who so desperately needed his response. Oddly, our seminarian was not too far from Kant (1797/1993), who once argued in favor of telling the truth to a murderer regarding the whereabouts of his intended victim. A deceit, he noted, could make one an incidental accomplice, whereas the truth could never permit so much. In both cases the need for preserving the moral purity of the subject is at issue. How unfortunate for the world . . . and for ethics! Sometimes it's best to feed the hungry for the wrong reasons rather than not at all. Sometimes it may be best to deceive.

Acknowledgments

This chapter follows in part from a presidential address ("Is Ethics Computable, or What Other Than *Can* Does *Ought* Imply?") presented at the 2011 annual meeting of the International Association for Computing and Philosophy, held at Aarhus University in Denmark, July 4–6, and in part from a conference presentation ("Between Angels and Animals: The Question of Robot Ethics, or Is Kantian Moral Agency Desirable?") presented at the 18th annual meeting of the Association for Practical and Professional Ethics, held in Cincinnati, Ohio, March 5–8, 2009, though it has other predecessors as well. In this regard, I wish to thank the following people for their comments and conversations that led up to this paper: Colin Allen, Susan Anderson, Selmer Bringsjord, Larry Colter, Richard Connolly, Charles Ess, Luciano Floridi, Christopher Harrison, Deborah Johnson, Patrick Lin, Dianne Oliver, James Moor, Matthias Scheutz, and Wendell Wallach.

References

Allen, C., Varner, G., & Zinser, J. (2000). Prolegomena to any future artificial moral agent. *Journal of Experimental and Theoretical Artificial Intelligence, 12,* 251–261.

Anderson, M., & Anderson, S. (2007). Machine ethics: Creating an ethical intelligent agent. *AI Magazine, 28*(4), 15–26.

Beavers, A. (2001). Kant and the problem of ethical metaphysics. In M. New, R. Bernasconi, & R. Cohen (Eds.), *In proximity: Emmanuel Levinas and the eighteenth century* (pp. 285–302). Lubbock, TX: Texas Tech University Press.

Beavers, A. (2009, March). *Between angels and animals: The question of robot ethics, or is Kantian moral agency desirable?* Paper presented at the 18th annual meeting of the Association for Practical and Professional Ethics, Cincinnati, Ohio.

Beavers, A. (2010). Editorial to *Robot ethics and human ethics* [Special issue]. *Ethics and Information Technology, 12*(3), 207–208.

Beavers, A. (2011). Moral machines and the threat of ethical nihilism. In P. Lin, G. Bekey, & K. Abney (Eds.), *Robot ethics: The ethical and social implication of robotics* (pp. 333–344). Cambridge, MA: MIT Press.

Connolly, R. (2011, September). *Are corporations moral agents?* Paper presented at the Andiron Lecture Series, University of Evansville, Indiana.

Dennett, D. (2006, May). *Computers as prostheses for the imagination.* Paper presented at The International Computers and Philosophy Conference, Laval, France.

Floridi, L. (1999). Information ethics: On the philosophical foundation of computer ethics. *Ethics and Information Technology, 1,* 37–56.

Floridi, L. (2002). On the intrinsic value of information objects and the infosphere. *Ethics and Information Technology, 4,* 287–304.

Floridi, L., & Sanders, J. (2001). Artificial evil and the foundations of computer ethics. *Ethics and Information Technology,* 3, 55–66.

Floridi, L., & Sanders, J. (2004). On the morality of artificial agents. *Minds and Machines, 14*(3), 349–379.

Kant, I. (1993). *Grounding for the metaphysics of morals* (J. Ellington, Ed. & Trans.). Indianapolis, IN: Hackett. (Original work published 1785)

Kant, I. (1993). On a supposed right to lie because of philanthropic concerns. In J. Ellington (Ed. & Trans.), *Immanuel Kant: Grounding for the metaphysics of morals; with, on a supposed right to lie because of philanthropic concerns* (pp. 63–68). Indianapolis, IN: Hackett. (Original work published 1797)

Kurzweil, R. (1990). *The age of intelligent machines.* Cambridge, MA: MIT Press.

Levinas, E. (1969). *Totality and infinity: An essay on exteriority* (A. Lingis, Trans.). Pittsburgh, PA: Duquesne University Press. (Original work published 1961)

Levinas, E. (1989). The pact. In S. Hand (Ed.), *The Levinas reader* (pp. 211–226). Cambridge, MA: Blackwell. (Original work published 1982)

Levinas, E. (1990). And God created women. In A. Aronowicz (Trans.), *Nine Talmudic readings* (pp. 161–177). Bloomington: Indiana University Press. (Original work published 1977)

Moor, J. (2006). The nature, importance and difficulty of machine ethics. *IEEE Intelligent Systems, 1541–1672,* 18–21.

Royakkers, L., & van Est, R. (2010). The cubicle warrior: The marionette of digitized warfare. *Ethics and Information Technology, 12*(3), 289–296.

Turing, A. (1950). Computing machinery and intelligence. *Mind, 59,* 433–460.

CHAPTER THIRTEEN

Instrumental Play or the Moral Risks of Gamification

Miguel Sicart

The world of technological development and computer culture seems to be driven by the perpetual motion of hype machines. From the dot-com era to the current focus on green technologies, the Silicon Valleys of the world fuel, and are fueled by, our hopes and dreams, our ambitions and aspirations.

The evangelists of these cultures tend to paint a future of technological pervasiveness that might remind us of classic dystopian fictions, worlds in which corporations know every detail about your taste, in which free services mask trading of identities and information, in which life without computers is unthinkable.

The later years of the aughts saw the so-called gamification paint a future in appearance less bleak and more playful than that of other technological promises. The gamified future took the world by storm with the promise of saving it by making people play. Game designers and marketers poured their efforts in convincing us, consumers, that adding a game layer to the world will not only fix it but also improve everybody's economy.

Needless to say, I can be counted among the skeptical. My concern is that of an ethicist interested in games, and my question is a simple one: what are the moral concerns that emerge when "ludifying" reality?

In this chapter, I want to explore some moral concerns raised by the phenomenon of gamification. My goal is to describe and analyze some of the ethical risks that this type of product design and marketing might have but also to provide a number of arguments and approaches that would keep the positive elements of gamification still present and relevant, while ameliorating the potential concerns.

For this purpose, I have chosen a multidisciplinary approach. The philosophical background for my analysis of gamification is double: Adorno's (2001) critical theory will be the anchoring point for the reflection on gamification and instrumentality; Floridi's (2005, 2010) information ethics and philosophy of information, and particularly his constructivist anthropology, will not only analyze why gamification is morally risky but also how it can become morally beneficial. These two main theoretical discourses will be supplemented by a close reading of relevant Human Computer Interaction literature (Gaver, 2009; & Sengers & Gaver, 2006; Hallnäs & Redström, 2001; Sengers, Boehner, Shay, & Kaye, 2005).

The more ambitious goal of this chapter is to reconnect gamification (understood as Deterding, Kahled, Nacke, Dixon[2011]) with a larger, more important tradition. Like many new cultural fevers, gamification seems to come up from nowhere and discover a promised land no one had ever thought of before. The argument in favor of gamification, or more broadly, my praise of the ludic turn (Raessens, 2006)—understood as the cultural focus on the ludic aspects of life and culture—will be based not on games, or design, but on the fascination that play and playing have, which has captivated philosophers, artists, and artisans throughout history (DeKoven, 2002; Suits, 2005; Sutton-Smith, 1997). This chapter wants to encourage scholars to think about playfulness rather than gameness (Deterding. Kahled, Nacke, Dixon, 2011) as the source for the compelling, educative, creative, and emotionally satisfying experiences that gamification promises. I will sketch a theory of play for the ludic turn, raising the ante on gamification and challenging what has become a standard way of thinking about the importance of games in culture.

Gamification and the Ludic turn

Even though the concept of gamification is being popularized on vague pretenses of scientificism (Zichermann & Cunningham, 2011; Zichermann & Linder, 2010) and grandiose rhetorics of world saving might (McGonigal, 2010), the reality behind the concept is deceptively simple.[1] Gamification, a term used to describe the application of game-based reward structures to products and services designed for measuring, tracking and occasionally teaching about repetitive, or repeatable tasks, proposed a more scientific and appropriate definition:

> "Gamification" refers to
> - *the use* (rather than the extension) *of*
> - *design* (rather than game-based technology or other game-related practices)

- *elements* (rather than full-fledged games)
- *characteristic for games* (rather than play or playfulness)
- *in non-game contexts* (regardless of specific usage intentions, contexts, or media of implementation). (p. 5)

This definition breaks down the often-repeated idea that gamification consists of adding a game layer to the world. Gamification advocates (Zichermann & Cunningham, 2011) appeal to data and incidental evidence that point to the notion that games, due to their design, not only engage people in mundane and repetitive tasks, but this engagement is somewhat correlative to a learning process, and therefore, games can be used for teaching and learning purposes.

However, gamification advocates do not seem to be able to bridge the gap between general aspects of game design, such as context dependence (Wilson, 2011) player psychology (Canossa & Drachen, 2009; Fullerton, 2008; Schell, 2008), gameplay progression, or even more arcane topics such as game feel (Swink, 2009). It seems like gamification does not actually involve many of the elements that make game design a modern craft. There is none of the advanced user experience tools and theories that are now so popular in game design, such as artificial intelligence-driven player modeling. Gamification design has focused on the most simple, superficial, and likely less satisfying tool of game design: reinforcement structures and rewards through positive feedback loops.

A quick observation of the most popular, and cited, gamification tools (Zichermann & Cunningham, 2011) shows an abundance of badges, progression ladders, points, and score lists. One could think that gamified systems and services have taken their design cues from the age of arcade machines rather than from the modern craft of game design. Of course, there is a novelty in it, one that justifies paying more careful attention to gamification than what the current state of affairs might encourage us to. Gamification excels at adapting to new contexts of use, at finding services and situations that may actually benefit from being partially designed as games (Korhonen, Montola, & Arrasvuori, 2009; Morrison, Mitchell, & Brereton, 2007.

In other words, while gamification as a current phenomenon has only taken the most superficial and less interesting elements of game design, such as badges and points, it has also shown how the work-obsessed Western culture may actually require a look back at play and play forms to renew itself (DeKoven, 2002; Suits, 2005). Gamification, as a design phenomenon, is insignificant and unimportant. However, as a snapshot of our zeitgeist, gamification is extremely interesting. For all its ethical maladies, gamification is a symptom of the ludic turn in Western culture.

This ludic turn can be described as a trend by which play, playfulness, toys, and games are taking a central part in our culture, replacing old forms of entertainment and spreading its influence beyond the limits of leisure. This is perhaps the most interesting aspect of the ludic turn—it is not about forms of leisure but about extending the attitudes, tools, and techniques of the ludic sphere to the world of work and production.[2] Play and playfulness are not anymore confined to the limits of games and free time. Work becomes more playful, tools become closer to toys, and productivity can and is measured in terms of points and scores, rather than dry pie charts.

The ludic turn, of course, has strong cultural roots in twentieth-century arts and philosophy and, arguably, in a subterranean yet highly influential stream of thought that can be traced back to Ancient Greece. Huizinga's *Homo Ludens* (1992), the work that sits at the center of the ludic turn's canon, argued that play is at the center of culture as a source of arts, religion, thought, and emotions. Caillois (2001) and Sutton-Smith (1997) explained why games, as the privileged expression of play in culture, are more than just childish pastimes, making them genuine forms of expression and representation of the world.

Games and play are not the only cultural elements fueling this ludic turn. The popularity of festivals like the Come Out & Play Festival (http://www.comeoutandplay.org/) points at the resurgence of the carnivalesque. Bakhtin's (1984, 2008) theory of the carnival can explain why this playful approach to public spaces bears more importance than just that of the purely festive: by making the festive a central part of the expression of a culture, order and authority are subverted, and a certain sense of appropriative play, a playful attitude, becomes a governing force of expression.

The importance of play and playfulness, the core of this ludic turn, has also permeated the discipline of design. Even though I address that literature more closely later, this brief account of the ludic turn requires mentioning how design recognizes the importance of play in culture. We often acknowledge that good design is functional, that objects should be created to perform their functions efficiently and as unobtrusively as possible.

Current trends in design research take a different stand. Instead of priming efficiency and unobtrusiveness, maximum importance is given to the user, his or her context, and his or her creativity and emotional well-being. Design is not anymore about an object that solves a task but about an object in an environment that, together with a user, performs a task to which it is not necessarily neutral (Gaver, 2009; Sengers & Gaver, 2006; Sengers et al., 2005).

However, not all is bright in this ludic turn. The virtues of labor and efficiency are not to be disdained, and the particular nature of play, always volatile, with its capacity to create but also to enjoy destruction (Schechner, 1988), may raise ethical concerns. A move toward the ludic aspect in culture is, in the light of the classic philosophy of play (Esposito, 1988; Fink, 1988), a positive thing, inasmuch as the limits of play are acknowledged. Play was defined as an autotelic activity excluded from morality (Huizinga, 1992). Therefore, in a culture of play, how can we make sure that we are not fooled by play, let into play, and played by the very same services and entertainment that we are enjoying? How can we practice our ethics through play, if there are no ethics in play?

Play is dangerous. Play is a way of being in and understanding the world. But it is also an attitude that may lead to disdain, foolishness, superficiality, and actions in which the means to their ends are not contemplated.[3] The ludic turn is an exciting moment in culture, but it demands moral reflexes and solid thinking.

In this sense, gamification is an example of the moral perils of the ludic turn. Gamification has misunderstood play. By giving us a world of games, gamification might deprive us of the pleasures of the anarchic play, of the appropriative nature of play. Gamification, as a phenomenon, is not focusing on the positive, creative aspects of play but on the superficial, instrumental tools that make the outer skin of games: badges, goals, achievements. Gamification makes services, activities, and even social causes look and feel like games, yet they are not necessarily playful, or at least they are not embracing the properties of play as a way of expression and understanding. In that lack of playfulness lies the moral risk of gamification. In the next sections I explain this argument.

The ludic turn, as said, can be a turn toward the carnivalesque, creative, protean play that has been nurturing our culture. However, gamification, at least understood in the fashionable sense noted at the beginning of this section, has focused exclusively in the means to play, rather than in play itself. As such, it has turned play and playfulness into autotelic labor, and that is what makes it an instrumental tool that should morally concern us.

A Portable Theory of Play

Why do we play? Is it for the indefinable pleasures of fun, to run away from the dull routines of daily activities, or because we ethologically *need* to play? Or maybe it is because, as Huizinga first argued, it is an integral part of our culture and, incidentally, of our ways of expressing and understanding truths about the world.

However, if we look closer at what the activity of play is about, we might initially agree with the classic argument that play is an autotelic activity, an activity that is valuable in and of itself, and that it has no goals, or purposes, outside the activity as such. Of course, autotelic just means that the activity has no external goals, not that it doesn't affect whatever stays outside of the sphere of the game. Games are not played in an airtight magic circle (Consalvo, 2007, 2009). Games, and play, do not occur in an enclosed space detached from the world external to play. In fact, they happen in a conversation, a back and forth between whom we are in play and who we are outside of play (Sicart, 2009; Stenros & Waern, 2011). That's why we can conceive play as a form of expression, of education, and of communication—because it is enclosed but not isolated.

However, what I am interested in exploring here is the concept of play as an autotelic activity and what implications it has for understanding the ethical perils of gamification. Even though this is not the place to write a comprehensive theory of play, I would like to sketch a brief understanding of play, a microtheory of play for the ludic turn. I will use this theory of play to explore the implications of instrumentality in the design and deployment of gamified systems and why they pose more ethical threats than what might be expected.

The brief theory of play I am outlining here describes play as an activity that is autotelic, creative, appropriative, and disruptive, and above all, highly personal. Let me briefly characterize each of these.

Play is autotelic, in that it is an activity with its own goals and purposes, with its own marked duration and spaces, and with its own conditions for ending. The boundaries of autotelic play are not formally rigid. Play is autotelic in a context (Dourish, 2004), and it is an activity people engage with. Therefore, context and communities will play a role in the constant negotiation of the purpose of play (Taylor, 2009). Play has a purpose of its own, but that's not a purpose that can be fixated. Play activities can be described as diachronically or synchronically autotelic, that is, looking at how the purpose of play evolved though the play session or looking at what particular purpose a particular instance of play had in a particular session. Descriptions of play should take these two perspectives into account as well as the context and participants in play.

Play is creative, in that it affords from players different degrees of creativity that are inherent to the play activity itself. Playing is accepting the rules of the game and performing within them according to our needs, personality, and constitution of the playing community we're engaged with. Play is the act of creatively engaging with the world, with technologies, contexts, and objects, from games to workplaces, trying to explore them in a ludic fashion. Play creates their objects, and their communities, and thus is a creative force.

Play is appropriative, in that it takes over the context in which it is applied and cannot be totally predetermined, or predicted, by such context. Playing a game is more than just understanding and mastering a system—it is the act of appropriating that system, turning it into an activity in which it is players who look at the context in which they play and behave accordingly to that context.

Play is disruptive as a consequence of being appropriate. Play, when taking over the context in which it takes place, breaks a state of affairs. This is often done for the sake of fun, enjoyment, for passing pleasures. But like all passing pleasures, play can also disruptively reveal our conventions, assumptions, biases, and dislikes (Stenros & Montola, 2011) In disrupting the normal state of affairs by being playful, we can go beyond fun—we appropriate a context with the intention of playing with and within it. And in that move we are revealed, and reveal, the inner workings of the context we inhabit.

Play is personal. Even though we play with others, the effects of play are always individual (Waern, 2011), attached and wielded to our sentimental, moral, and political memories. Who we are is also who plays, the kind of person we let loose when we play. Our memories are composed of these instances of play, the victories and defeats but also the shared moments. Play is not isolated in our eventful lives—in fact, it is a string with which we tie our memories together, our friendships; it is a trace of character that defines us.

This tiny theory of play invokes a very specific cultural understanding of play. The idea of play as disruptive, creative, and appropriative is tightly coupled with Bakhtin's notion of the carnivalesque. In fact, this idea of play should be seen as a carnivalesque play, one that does not necessarily happen in the open spaces of the Middle Ages, but in the social networks, implicit or explicit, within which we play games. Much like Bakhtin's carnival, play and playfulness is a movement that puts human agency at the center of action, standing against disciplinary systems of control, from games to strict servitude to the corporation. Play creates and disrupts, and in doing so, it becomes a carnival.

The other theoretical framework that contributes to this understanding of play is Schechner's (1988, 2006) concept of "dark play," as related to performance arts. While play is acknowledged to be a fundamental element in the performance arts domain, Schechner, in his own interpretation of what play is, evokes a type of play that is fundamentally different from the childish, positive play that we know about from play theory. What Schechner proposed is the understanding of a type of play that is challenging, exploring boundaries and what is acceptable and what is not, in fact, a type of play that explores what play is.

Dark players are often concealing either the fact that they are themselves playing or that the activity they are engaging with is a game. Dark play is dangerous, exploratory, an extreme of the activity of play. But it also illustrates an approach to play and playfulness that is particularly enriching: play as a critical activity and playfulness as a critical attitude toward the world.

Given this understanding of play, why does gamification fail? If we look at gamification from the autotelic perspective, the failure is obvious: gamification advocates for play not as an activity with its own goals but as an instrumentalized activity for other purposes. This leads to a complete breakdown of play for gamification: it cannot be creative or disruptive, as it is an activity determined, designed for, and applied to *other goals than playing*. Play has become instrumentalized.

Instrumentality

Instrumental play is a concept that adapts the theories of Adorno and Horkheimer (2010) with the purpose of explaining the limits of understanding game design as a tool that generates the means for ends other than play. Instrumental play is a take on the concept of instrumental rationality first written by the Frankfurt School of Critical Theory in the 1950s. Essentially, instrumental rationality is an argument used in the critique of modernity and its faith on a model of reason that, despite its focus on proof and evidence, is still deeply rooted in myth. The fundamental tension lies between myth and reason, and the excessive focus of enlightenment on reason as ends for the means of the modern revolution. In other words, reason becomes an end, a justification for all actions, and what enlightenment does is to rationalize all processes in order to justify them, even if those processes are deeply rooted in myth, in the irrational, in those aspects outside of the frames of rationality.

In gamification paradigms, play has a meaning, and a purpose, external to the activity of play, a goal that is decided prior to the design of the game and the performance of the activity: I do the chores because I earn points. As quoted in *The Dialectic of Enlightenment,* "For enlightenment is as totalitarian as any system . . . for enlightenment the process is always decided from the start" (Adorno & Horkheimer, 2010, p. 24). Much like enlightenment, then, gamification thinking is a determinist, perhaps even totalitarian, approach to play, an approach that defines the action prior to its existence and denies the importance of anything that was not determined before the act of play.

The problem with this understanding of reason is that, in critical theory terms, it is used to substitute myth and ritual in our culture, becoming the center of cultural and economic modernity:

> The technical process, into which the subject has objectified itself after being removed from the consciousness, is free of the ambiguity of mythic thought as of all meaning altogether, because reason itself has become the mere instrument of the all-inclusive economic apparatus. (Adorno & Horkheimer, 2010, p. 30)

But play belongs to the myth domain (Gadamer, 2004; Huizinga, 1992), to the area of rituals as much as to the domain of reason. If modernity despises ritualism, then games and play should shy away from what cannot be reasoned.

This leads to an understanding of play, and leisure, as mechanical outcomes of processes, outcomes that follow the same production and consumption models as labor:

> Amusement under late capitalism is the prolongation of work . . . mechanization has such power over a man's leisure and happiness, and so profoundly determines the manufacture of amusement goods, that his experiences are inevitably afterimages of the work process itself. (Adorno & Horkheimer, 2010, p. 137)

Play becomes external to the player and the play context.

Games create, frame, and encourage play. But what is the type of play games are created for? Caillois is defined it as "a free and voluntary activity, a source of joy and amusement . . . play is a separate occupation, carefully isolated from the rest of life, and generally is engaged in with precise limits of time and place" (Adorno & Horkheimer, 2010, p. 6). Again, this definition of play is very much engaged with the nature of games. But there is more to play than just this. Eugen Fink (1988) gave play a different meaning, a metaphysical one: "From the beginning, play is a symbolic act of representation, in which human life interprets itself" (p. 107). Again, play is more than just engaging with the rules of a game; it is a creative, productive experience: "Human play is a creation through the medium of pleasure of a world of imaginary activity" (Fink, 1988, p. 106).

Games structure play and facilitate it by means of rules. This is not to say that rules determine play: they focus it, they frame it, but they are still subject to the very act of play. Play, again, is an act of appropriation of the game by players. The discourse of gamification is dangerously close to that of scientificism, focused on the benefits that games, understood as a system of extrinsic motivation, can have on their players. In this sense, gamification advocates for a type of play that is instrumentally guided toward the completion of goals that ensure, by means of the objectively and scientifically solid procedural elements of the game, particular and effective results in the personal development of the player.

Gamification presents an argument that ties technology, systems, and reason together, justifying the existence of games as a tool for world-saving and personal improvement. However, this is achieved by means of ignoring players and play as a form of expression and exploration, as a way of experiencing values as much as adopting values. Instrumental play exists, and on occasion, it is useful to understand particular behaviors of players. But justifying the aesthetic, political, and ethical capacities of games by means of instrumentality leaves aside the complexities and nuances of play as appropriation and of players as cocreators of the ludic experience.

The moral risk of gamification, then, is closely connected to this interpretation of its proposals as an instrumentalization of play. What gamification aspires to is not to provide reasons to help human beings morally flourish (Bynum, 2006). In fact, gamification does not encourage the type of personal reflection that is required for an edifying activity to be morally approvable. What gamification does is externalize moral reasoning, so "players" don't need to exercise their moral thinking. Somewhere, somebody has already encoded a system of values that is communicated through a system of rewards, so users don't need to exert moral thinking.

In fact, I would argue that gamification sanitizes play—it strips it of all kinds of reflective, critical, appropriative, and provocative attitudes and translates it into a simple system of reward and consequence, a set of procedures. Play is dangerous and creative but also destructive. Gamification sanitizes play in order to allegedly tap on the benefits of play—but by doing so, it deprives its users of the possibility of using play as a moral instrument to engage with the world, as a creative tool to perform and enact who they are and who they want to be.

Again, the allure of gamification resides in its alleged taming of the risks of play in favor of structured rewards that lead to improved behaviors and a superficial bettering of the world. However, that is an instrumental take on play, one that denies the creativity of the performance of play. Is it possible to find examples of the design of objects that try to cater for playfulness, without falling into the sanitization of play? In the next section, I review some of the theories that support an alternative way of thinking on the use of play and playfulness in service design and how they can support moral development in a more complete way.

Playful Interaction Design

Design is about more than just the manufacturing of objects or services and much more than just about the creation of uses and prediction of behaviors. Design is the discipline that studies how objects are created and how designers can imbue their visions in their technological interpretations of the users' needs (Cross, 2007). Or, in other words, design is also about the ideas in the objects (Latour & Akrich, 1992; Winner, 1986).

Design is also the discipline that allows us to look and analyze how objects operate and why particular decisions that affect use were taken. By looking at how objects are designed and understanding how design operates as a discipline, it is possible to make arguments about the use of objects and the way they affect their users' lives.

Historically, design has been concerned with how objects are created for particular uses or needs, focusing on how forms can best follow function (Doordan, 1995). Since the inception of design as a discipline, designers have been concerned with the world of the artificial (Simon, 1996), with technologies and objects and their properties. The advent of digital technologies only increased that focus.

The work of Donald Norman, however, shows the transitions that design theories suffered in the last decades. Norman's intellectual trajectory shows how design theorists stopped focusing exclusively on form and functionality and added emotion to usability. It was a return to the intangible values of the artificial, to the fact that while efficiency is a merit in its own, so is the capacity of an object designed to be beautiful to move us. Design is, then, more than just about finding the most optimal function but also about arousing the right emotions, invoking the appropriate values in the user.

This slow but determined shift toward more human-centered and human-centric design opened up for the heavy influence in design of theories like activity theory or phenomenology (McCarthy & Wright, 2004; Mitcham, 1994; Verbeek, 2006). The focus was displaced from the user and the function to the context and the activity; being an appropriate design is one that is functional in a particular context for a particular set of users but also one that embodies the needs and values of the core group of users.

In recent years, design research has started looking elsewhere than the machines and objects themselves and paying more attention to the role of context, social situation, and mediation. It is not only the rise of ubiquitous computing (Dourish, 2004)—but also the observation that design and technology are more evidently moving into being a part of a context, and they are not anymore servants of humans but perhaps our coexistants. Relatively recent philosophical trends in design thinking have led to the consideration of technologies as more than mediators but as ways of embodying modes of understanding and existence, ontologies rather than just instruments (Ihde, 1990; Verbeek, 2006). A more radical turn, that of the *Philosophy of Information* (Floridi, 2010), is actually advocating for an ontology in which the materiality of technology might be a secondary concern, as ontology is not any more defined by biology or hardware but by information.

In this turn toward the object, the so-called critical design approach is the most appropriate way of tying together the importance of play with the discipline of interaction design. Critical design sits at the intersection of design and art, with the intention of criticizing the instrumentalization of design and the excessive focus on user-friendliness (Dunne, 2006). According to these Frankfurt-school-inspired designers, artists, and theorists, those objects created by applying contemporary theories of design did not foster reflection or contribute to the development of the people using them. Critical design aspires to challenge the conventions of user-centered design by imbuing objects with the possibility of critical use and reflection—by making them, in the words of Dunne (2006), "user-unfriendly."

Design, then, is not any more servant to function but a mode of reflection, a way of accessing the world, with objects that are not any more designed for a purpose but for a reflective practice.

Two other conceptual approaches are useful for understanding the turn toward the ludic in modern design theories. First is the idea of slow technologies (Hällnas & Redström, 2001), for which, if technologies are to play a role in our political and ethical configuration, the myth of functionality needs to be dispelled. Technologies designed to foster our critical sense should not be any more fast and efficient, since those values obscure and occasionally void the reflective capacities of the user. A slow technology would be a response to this situation: an object, or technologically driven context, in which the designed *thing* itself makes the process of its functionality visible by being slow. No more will a technology hide its inner workings—it will show them, both diachronically and synchronically. A slow technology frees time for the user to think about the actions he or she is asking a particular object to perform and the impact of those actions in the context of use. That free time, Hällnas & Redström argued, might be used to critically reflect about the object, its performance, and its role in the world.

The second paradigm that is interesting with regard to the design of technologies for playful purposes is that of technologies staying open to interpretation. In classic, mental-model-driven design theories (Norman, 2002), the designer communicates his or her vision of how the object should be used in a particular context via the creation of a system image that is then interpreted by the user, who in turn creates a user image. The goal of classic design is to eliminate cognitive friction (Cooper, 2004), to make the system image as close to a designer image as possible, so the user can seamlessly learn to use the object for the purpose it was created.

However, some theorists (Sengers et al., 2005; Sengers & Gaver, 2006) have posed a challenging vision: what if the objects created were, by design, open to interpretation? What if designers create objects that are only complete, and meaningful, when users *appropriate* them in new contexts, to which the designed objects do not resist. This should not be done at the expense of usability or of functionality, but it would still be the design for a process of appropriation, for an openness toward the user and his or her capacities of interpreting the object. Designers and users, then, would stand linked by an object that, reducing system authority and allowing for multiple means of interaction, opens up to the possible emergent uses that depend on context and sociocultural situations in which technologies are deployed.

Bill Gaver (2009) tied together play and design research, pointing at the possibility of an interpretation of how to create objects for the ludic turn. Gaver argued that design should move toward more playful approaches, toward a better respect and understanding of play as one of the driving forces of human culture. Unlike rationality, play has interesting creative and appropriative capacities, and the design for play can effectively tap into those capacities and create objects that are not only playful but also thoughtful and respectful with the capacities of man to engage himself with his context.

Designing for homo ludens, then, invokes a rhetoric close to that of the theory of play outlined here: play, and objects designed for play, are not mere entertainment devices—they are ways of understanding and being in the world, of establishing relations, of creating politics, and of engaging with others.

Is gamification a type of design for the homo ludens or critical design? Given its focus on instrumentality, and its lack of some of the characteristics that define what Gaver and also Sengers and Hällnas and Redström have theorized, the answer is negative. Gamification is not a design for the type of creative, critical play that engages users in new modes of understanding the world. It is, in fact, quite the opposite: it is the design of external rewards for the functional completion of activities, regardless of the meaning of those activities. Ludic design is not the design of game-alike systems on top of real behaviors but the design of services and products with an idea of creative play.

However, even though ludic design shows a way of approaching the creation of products and services that cater for a creative, appropriative type of play in consonance to the theory of play here outlined, there is still the lingering question of values. If gamification is morally risky because of its instrumentalization of behaviors, how can ludic design avoid that risk? How can it escape the fate of the Huizinga-defined homo ludens, which claims that play is outside morality?

Ethics Through Play

The design of ethical experiences for homo ludens begs for an ethical theory that allows for a creative activity mediated by (digital) technologies that both constrain and expand agency. In fact, what this chapter proposes is an even more radical approach, more than adjusting an ethical theory to the particular requirements of the design of play. What is proposed is a theory of (play) design that conforms to a philosophical theory and, therefore, to an ethical theory.

Thus, I sketch an informational theory of play design, based on Floridi's (2010) *Philosophy of Information*. The main contributions of the Philosophy of Information for this chapter are its ethical theory, information ethics, and more specifically, its constructivist approach to agency in informational environments (Floridi & Sanders, 2005). What this section does is describe a balancing act between the technical rigor of the PI and the need to expand this understanding of philosophy toward the domain of play and the playful in order to answer the ethical questions raised by gamification but also to inspire new forms of approaching the design of playfulness.

Let's start with information ethics, as defined by Floridi (2010). This ethical framework is based on the Philosophy of Information. In Floridi's terms, the Philosophy of Information is "the philosophical field concerned with (a) the critical investigation of the conceptual nature and basic principles of information, including its dynamics, utilization, and sciences, and (b) the elaboration and application of information-theoretic and computational methodologies to philosophical problems" (p. 14). This focus on information provides the Philosophy of Information with

> one of the most powerful conceptual vocabularies ever devised in philosophy . . . because we can still rely on information concepts whenever a complete understanding of some series of events is unavailable or unnecessary for providing an explanation. In philosophy, this means that virtually any issue can be rephrased in informational terms. (Floridi, 2010, p. 16)

A key concept in Floridi's philosophy is informational agency, which extends beyond anthropocentric and biocentric approaches, to include any type of relevant informational agent. This definition of agency allows for the inclusion of artificial agents in the ontological domain, including software or adaptive software systems (Floridi & Sanders, 2005).

The method of the Philosophy of Information, the method of abstraction, is based on object-oriented programming concepts. To understand the ontology of information, agents and patients should be treated as informational objects with methods, properties, and interactions (Floridi, 2010). In terms of analysis, a philosophical analysis has to be approached from a certain *level of abstraction* (Floridi, 2010, pp. 48–58).The most relevant outcome of this informational approach has been the formalization of an information ethics, an "*ontocentric, patient-oriented, ecological* macroethics" (Floridi & Sanders, 2005, p. 10). Since it is based on the informational ontology of the Philosophy of Information,

> the ethical discourse now comes to concern information as such, that is not just all persons, their cultivation, well-being and social interactions, not just animals, plants and their proper natural life, but also anything that exists, from paintings and books to stars and stones; anything that may or will exist like future generations; and anything that was but is no more, like our ancestors. (Floridi, 1999, p. 43)

Also, information ethics takes a clear constructivist approach: "Ethics is not only a question of dealing morally well with a given world. It is also a question of constructing the world, improving its nature and shaping its development in the right way" (Floridi & Sanders, 2005, p. 2).

Information ethics is a very abstract and somewhat verbose ethical theory. Its main strengths lie on the strong methodology that allows for the ethical scrutiny of agents, technologies, and patients in the context of information systems.

What is important for our ethical understanding of play and playful design is not only the abstract concepts that form the theoretical underpinnings of information ethics—what is fundamental is understanding who is an ethical agent in this perspective and how the agent can behave or be addressed ethically. The basic tenet of information ethics' ethical anthropology is the idea of an agent as a creative steward of the environment in which he or she has agency—an agent that needs to be able to protect as much as to contribute to the richness of the informational environment.

Before describing with more depth this constructionist approach to the ethical being in an informational environment, it is necessary to briefly look back to one of the fundamental concepts in playful interaction design. The ideas of designing for homo ludens, of creating objects open for interpretation, of slow technologies, all point at play as a creative, appropriative activity, one that is rich and valuable because it is not tied to functional, rational, instrumental thinking. Designers of playful devices engage with creative forces with the excuse, or through the instability, of the object. That instability is actually a source of responsibility and values, the space in which politics and ethics can thrive through play.

That openness connects with the general constructivist idea of information ethics' anthropology. The idea of constructivist ethics is a rather classic one. From Aristotelian virtue ethics to Floridi's information ethics, constructivist ethics see the capacity of humans to develop values as a constant activity, as a process that leads, by exerting the best of our values, to fulfilling our potential. What information ethics argues for, actually, is an agent that should strive for flourishing not only by trying to fulfill the best of his or her potential but by being in the context and environment where he or she is an agent and by contributing to it. This is perhaps the most important aspect of information ethics' ethical agency concepts: agents, in order to fulfill their potential, need not only to perform actions but also to creatively participate in the very constitution of the experiences of which they are a part.

Floridi and Sanders (2005) named this type of agency "homo poieticus," the creative being that appropriates the environments with respect in order to flourish, to develop the best of his or her capacities. The essential ethical move of the homo poieticus in the context of the information ethics framework is the so-called creative stewardship, the modality of being by which an agent is not any more demanded to make the environment healthy but is also ethically required to creatively contribute to it.

Homo poieticus makes the homo ludens accountable. Homo ludens, in the classical sense, was exclusively dedicated to play, and the creativity in play was a consequence rather than a requirement of the activity. Homo ludens played creatively because that is how play is. Homo poieticus places ethical value in the act of the creative appropriation. Creative stewardship can be used to harness the capacity of play to engage agents in creative, appropriative experience of play and turn that activity of play into an ethically relevant action.

This perspective allows for yet another critique of the instrumental design of gamified systems. Gamified systems are hardly open for creative stewardship—they are "just" input/output systems that may deliver, if seen externally, results in the modification of actions, but they do so only because agents perform the actions externally motivated, not because they engage with the world in a creative manner. Most gamification systems focus on the stewardship domain—values are provided by the designer or corporation behind the product. In doing so, the user is deprived of ethical agency and is at the mercy of a narrow system of rewards and challenges that may not map to a more appropriative domain.

What are the challenges, then, when trying to design for the homo poieticus? The ideas proposed by these interaction design theorists still hold: systems open to interpretation that downplay system authority are a key for designing for more playful interaction. Even highly politicized/ideologized design theories, like Dunne's "user-unfriendliness" or slow technologies, still point toward the same requirement: for a technology to be playful it needs to break with the dictates of functionalism and conventional usability.

The problem, then, might be of a philosophical rather than of a technical point of view. What the designing for homo ludens does not take into consideration that the agency model proposed by Huizinga under that name was strictly an *amoral* one, that of an agent involved in an activity devoid of moral or political meaning. The play theory on which the concept of homo ludens is functional is one at odds with strong ethical discourses.

What is needed, then, is a theory of play that matches with an understanding of ethical agency. The play theory sketched in this chapter, focused on the ideas of creativity and appropriation, is a good match with Floridi's idea of the homo poieticus, particularly with the idea of creative stewardship. The ethics of playful systems, from gamification to games to toys, is tightly coupled with a model of creative agency, with a homo poieticus.

In more concrete terms, the ethics of playful systems, both as objects and as activities, is that of the activity of play, or more specifically, the way in which it is designed to create a particular play experience. Ethics happens through play, or *in* play, facilitated by a system that acknowledges the user as a creative, appropriative agent, a homo poieticus who not only wants to play but also to preserve that experience of play and, if needed, interpret the world and himself or herself in the play activity.

It is, then, the performance of play that becomes a performance of ethics, and therefore, any system designed with the idea of exploring and enticing the ethical being of the user needs to be open for creative, appropriative play. If play is an interpretative activity, one in which leisure and entertainment become a way of performing an understanding of the environment in which we live, then the systems designed for play need to acknowledge that. It is no longer enough to design for the homo ludens—for an ethical agent that appropriates and creates as well as performs and follows instructions, the paradigm needs to change, and a design for the homo poieticus is necessary.

Toward Designing for the Homo Poieticus

Players are creative beings that take the context of a game, or of a situation, and engage with it by playing, by appropriating it in an autotelic, disruptive, personal performance that transcends instrumentality and looks for self-expression, for a tenuous balance between creation and destruction.

Classic play theories described this player as a homo ludens—yet in this chapter, a new paradigm has been suggested: the homo poieticus, the creative being that plays as a way of enacting his or her creative stewardship in the informational environment in which he or she exists. Being is, on occasion, expressed and enacted through play. Not, like in the case of many gamified systems, by means of adding external rewards to behaviors, but by affording playful behaviors, by enticing people to become playful—by turning us into players, openly or surreptitiously.

The title for this section takes a direct cue to Gaver's famous article on designing for homo ludens (2009). As mentioned in the previous section, the main problem with that design idea is how the original concept of homo ludens did not account for a moral domain. While homo poieticus does afford that moral domain, the question remains: are the paradigms proposed by design research still valid? Can we then say that there is a design for the homo poieticus?

The answer is yes. The playful interaction design research work still holds valid for an agent that is actively ethical. What these designers and theoreticians argued for was a model of technologies open for interpretation, playful technologies at the service of playfulness and openness, less concerned about the efficient fulfillment of a task and much more interested in becoming a partner in creative interaction with their users.

However, if the intention is to use play to engage people, agents, in ethical experiences and activities, if the purpose of the ludic turn is to extend play and play-alike activities to more "serious" domains than games, to more noble purposes than entertainment, then it is time to extend the domain of playful design toward the domain of morality.

Designing for the homo poieticus implies a new set of challenges for designers and philosophers: how to use play to contribute to human flourishing. Technologies should not be seen as means of rewarding or purposes in themselves. Technologies for ludic ethics need to be seen as an essential part of human flourishing, as the environments necessary, but not sufficient, for creative play to become a *practical ethical development.*

Play is a human expression that is returning to its central place in culture. Hidden in games and other forms of leisure, ludic activities are becoming more and more relevant in the way we express ourselves and explore the world. This ludic turn also calls for responsible design, one that does not trust in exploiting the instrumental pleasures of play, one that is aware of the risks of dark, excessive play. The ludic turn should not be exclusively focused on play as entertainment. Play is understanding and expression, values and creativity. By playing, we explore who we are. In the age of the ludic turn, it is time to design so that by playing, we also become better human beings.

Notes

1. For a history of the concept of gamification, see Deterding et al. (2011a).
2. This is, of course, not new (Flanagan, 2009; Huizinga, 1992; Korhonen et al., 2009; Morrison et al., 2007), but the popularity of the term gamification justifies this somewhat excessive argument, borrowed from Gaver (2009).
3. The idea of the corrupted play is present in Caillois (2001, Chapter IV).

References

Adorno, T. (2001). *The culture industry* (2nd ed.). New York, NY: Routledge.

Adorno, T., & Horkheimer, M. (2010). *The dialectic of enlightenment.* New York, NY: Verso.

Bakhtin, M. (1984). *Rabelais and his world.* Bloomington: Indiana University Press.

Bakhtin, M. (2009). *Problems of Dostoevsky's poetics* (C. Emerson, Trans.). Minneapolis: University of Minnesota Press.

Bakhtin, M. M. (2008). *The dialogic imagination* (M. Holquist, Ed.; M. Holquist & C. Emerson, Trans.). Austin: University of Texas Press.

Bynum, T. W. (2006). Flourishing ethics. *Ethics and Information Technology, 8*(4), 157–173.

Caillois, R. (2001). *Man, play and games.* Urbana: University of Illinois Press.

Canossa, A., & Drachen, A. (2009). Play-Personas: Behaviors and belief systems in user-centered game design. *Lecture Notes in Computer Science, 5727,* 510–523.

Consalvo, M. (2007). *Cheating. Gaining advantage in videogames.* Cambridge, MA: The MIT Press.

Consalvo, M. (2009). There is no magic circle. *Games and Culture, 4*(4), 408–417.

Cooper, A. (2004). *The inmates are running the asylum. Why high-tech products drive us crazy and how to restore the sanity.* Indianapolis, IN: Sams Publishing.

Cross, N. (2007). *Designerly ways of knowing* (paperback ed.). Basel, Switzerland: Birkhäuser.

DeKoven, B. (2002). *The well-played game.A playful path to wholeness.* Lincoln, NE: Writers Club Press.

Deterding, S, Dixon, D., Khaled, R.,Nacke, L. (2011a). From game design elements to gamefulness: Defining "gamification." Proceedings of the 15th International Academic MindTrek Conference Envisioning Future Media Environments, 2011. Tampere, Finland: ACM, pages: 9–15.

Deterding, S, Dixon, D., Khaled, R., Nacke, L. (2011b). *Gamification: Toward a definition. Design,* 12–15. Vancouver. Retrieved from http://gamification-research.org/wp-content/uploads/2011/04/02-Deterding-Khaled-Nacke-Dixon.pdf

Deterding, S., Khaled, R., Nacke, L., & Dixon, D. (2011). *Gamification: Toward a Definition.* Presented at the CHI 2011, Vancouver, BC.

Doordan, D. P. (Ed.). (1995). *Design history. An anthology.* Cambridge, MA: The MIT Press.

Dourish, P. (2004). What we talk about when we talk about context. *Personal and Ubiquitous Computing, 8*(1), 19–39.

Dunne, A. (2006). *Hertzian tales. Electronic products, aesthetic experience, and critical design.* Cambridge, MA: The MIT Press.

Esposito, J. L. (1988). Play and possibility. In W. J. Morgan & K. V. Meier (Eds.), *Philosophic inquiry in sport.* Champaign, IL: Human Kinetics, 114–118

Fink, E.(1988).The ontology of play. In W. J. Morgan & K. V. Meier (Eds.), *Philosophic inquiry in sport.* Champaign, IL: Human Kinetics, 100–109

Flanagan, M. (2009). *Critical play. Radical game design.* Cambridge, MA: The MIT Press.

Floridi, Luciano. (1999). Information ethics: On the philosophical foundation of computer ethics. *Ethics and Information Technology 1*(1): 37–56.

Floridi, L. (2003). On the intrinsic value of information objects and the infosphere. *Ethics and Information Technology, 4*(4), 287–304.

Floridi, L. (2010). *The philosophy of information.* Oxford, UK: Oxford University Press.

Floridi, L., & Sanders, J. W. (2005). Internet ethics: The constructionist values of homo poieticus. In R. J. Cavalier (Ed.), *The impact of the internet on our moral lives.* Albany: The State University of New York, 195–213.

Fullerton, T. (2008). *Game design workshop. A playcentric approach to creating innovative games* (2nd ed.). Amsterdam: Elsevier.

Gadamer, H-G. (2004). *Truth and method* (2nd ed.). New York, NY: Continuum.

Gaver, William W. (2009). Designing for homo ludens, still. In *(Re)searching the digital Bauhaus.* Edited by Thomas Binder, Jonas Löwgren and Lone Malmborg. London: Springer, pp. 163–178.

Hallnäs, L., & Redström, J. (2001). Slow technology—designing for reflection. *Personal and Ubiquitous Computing, 5*(3), 201–212.

Huizinga, J. (1992). *Homo ludens. A study of the play-element in culture.* Boston, MA: Beacon Press.

Ihde, Don. (1990). *Technology and the lifeworld: From garden to earth.* Bloomington: Indiana University Press.

Korhonen, H., Montola, M., & Arrasvuori, J. (2009). Understanding playful user experience through digital games. In *Proc. DPPI,* Université Technologie de Compiègne, 274–285.

Latour, B., & Akrich, M. (1992). A summary of a convenient vocabulary for the semiotics of human and nonhuman assemblies. In W. Bijker & J. Law (Eds.), *Shaping technology/building society*.Cambridge, MA: The MIT Press, 259–264

McCarthy, J., & Wright, P. (2004). *Technology as experience.* Cambridge, MA: The MIT Press.

McGonigal, J. (2010). *Reality is broken: Why games make us better and how they can change the world.* London, UK: Penguin.

Mitcham, C. (1994). *Thinking through technology. The path between engineering and philosophy.* Chicago, IL: University of Chicago Press.

Morrison, A., Mitchell, P., & Brereton, M. (2007). The lens of Ludic engagement: Evaluating participation in interactive art installations. In Lienhart, R. & Prasad, A. R. (Eds.). *Proceedings of the 15th international conference on Multimedia*, Association for Computing Machinery Inc (ACM), Germany, Augsburg, pp. 509–512.

Raessens, J. (2006). Playful identities, or the ludification of culture. *Games and Culture,1*(1), 52–57.

Schechner, R. (1988). Playing. *Play & Culture, 1*, 3–19.

Schechner, R. (2006). *Performance studies. An introduction* (2nd ed.). New York, NY: Routledge.

Schell, J. (2008). *The art of game design. A book of lenses.* Amsterdam, The Netherlands: Morgan Kaufmann.

Sengers, P., Boehner, K., Shay, D., and Kaye, J. (2005). Reflective design. In *Proceedings of the 4th decennial conference on critical computing: between sense and sensibility*, Olav W. Bertelsen, Niels Olof Bouvin, Peter G. Krogh, and Morten Kyng (Eds.). New York: ACM, 49–58. DOI=10.1145/1094562.1094569 http://doi.acm.org/10.1145/1094562.1094569

Sengers, P., & Gaver, B. (2006). Staying open to interpretation: Engaging multiple meanings in design and evaluation. In *Proceedings of the 6th Conference on Designing Interactive Systems.* pp. 99–108

Sicart, M. (2009). *The ethics of computer games.* Cambridge, MA: The MIT Press.

Simon, H. A. (1996). *The sciences of the artificial* (3rd ed.). Cambridge, MA: The MIT Press.

Stenros, J., & Montola, M. (Eds.) (2011). *Nordic LARP.* Stockholm, Sweden: FëaLivia..

Stenros, J., & Waern, A. (2011). Games as activity: Correcting the digital fallacy. In M. Evans (Ed.), *Videogame studies: Concepts, cultures and communication.* Oxford, UK: Inter-Disciplinary Press, 11–23.

Suits, B. (2005). *The grasshopper: Games, life and utopia.* Peterborough, Ontario, Canada: Broadview Press.

Sutton-Smith, B. (1997). *The ambiguity of play.* Cambridge, MA: Harvard University Press.

Swink, S. (2009). *Game feel. A game designer's guide to virtual sensation.* Amsterdam, The Netherlands: Morgan Kaufmann.

Taylor, T.L. (2009). The assemblage of play. *Games and Culture, 4*(4), 331–339.

Verbeek, P. P. (2006). *What things do: Philosophical reflections on technology, agency, and design.* University Park, PA: Penn State University Press.

Waern, A. (2011). I'm in love with someone who doesn't exist! Bleed in the context of a computer game. *Proceedings of DiGRA Nordic Experiencing Games: Games, Play, and Players.* Stockholm: University of Stockholm,.

Retrieved from http://www.digra.org/dl/display_html?chid=10343.00215.pdf

Wilson, Douglas. (2011). Brutally unfair tactics totally OK now: On self-effacing games and unachievements. *Game Studies 11*(1).

Winner, L. (1986). *The whale and the reactor. A search for limits in an age of high technology.* Chicago, IL: University of Chicago Press.

Zichermann, G., & Cunningham, C. (2011). *Gamification by design: Implementing game mechanics in web and mobile apps.* Sebastopol, CA: O'Reilly.

Zichermann, G., & Linder, J. (2010). *Game-Based marketing: Inspire customer loyalty through awards, challenges, and contests.* Hoboken, NJ: Wiley.

Author Biographies

Anthony Beavers is a professor of philosophy and director of Cognitive Science and the Digital Humanities Lab at the University of Evansville in southern Indiana. He currently serves as president of the International Association for Computing and Philosophy and has written several papers concerning how information and communication technology may impact our attitudes concerning morality.

Brian Carey is a PhD student with the Manchester Centre for Political Theory (MANCEPT) at the University of Manchester, UK. His doctoral research concerns the application of abstract normative principles to nonideal, "real-world" scenarios.

Katherine Carpenter is a JD/MA student at the University of Denver studying international law and global health. Ms. Carpenter has worked in bioethics and served on an Institutional Review Board at the University of Washington prior to graduate school. She hopes to make the world a little bit safer, more ethical, and better through scholarship and policy implementation.

Vanessa P. Dennen is an associate professor of Instructional Systems at Florida State University. Her research investigates the nexus of cognitive, motivational, and social elements in computer-mediated communication, concentrating on two major issues: learner participation in online activities and interactions, norm development, and informal learning within online communities of practice.

David Dittrich is a computer security researcher at the University of Washington with over 20 years of experience responding to computer attacks. He has published numerous articles involving the legality and ethics of computer security response going back to 2001 and has served as a member on one of the University of Washington's institutional review boards for more than 2 years.

Meghan Dougherty is an assistant professor of Digital Communication at Loyola University Chicago's School of Communication and director of the school's graduate program in Digital Media and Storytelling. She studies the preservation and interpretation of web culture, collaboration tools to aid knowledge production, and web archiving as an emerging cyberinfrastructure for e-research.

Alex Gekker is a recent honors graduate from Utrecht University's New Media and Digital Culture masters program. Currently an independent researcher, his background includes digital communications, journalism, and serious games.

Mark Grabowski teaches online journalism and media ethics at Adelphi University in Long Island. He previously worked as a lawyer and newspaper reporter.

Don Heider is the dean of the School of Communication at Loyola University Chicago and founder of the Center for Digital Ethics and Policy. He is an award-winning journalist and teacher who has done extensive research in virtual worlds.

David Kamerer serves as assistant professor in the School of Communication at Loyola University Chicago, where he teaches courses in public relations and new media. He earned a PhD in Telecommunications from Indiana University and is accredited in public relations (APR) by the Universal Accreditation Board.

Adrienne Massanari is an Assistant Professor in the Department of Communication at the University of Illinois—Chicago. Her research interests include gaming, new media, and digital ethics.

Jo Ann Oravec is an associate professor in the College of Business and Economics at the University of Wisconsin at Whitewater, with her MBA, MS, MA, and PhD from the University of Wisconsin at Madison. She taught one of the first "computers and society" courses in the 1980s (at UW-Madison), chaired the pioneering Privacy Council of the State of Wisconsin in the 1990s, and published *Virtual Individuals, Virtual Groups: Human Dimensions of Groupware and Computer Networking* (Cambridge University Press).

Erin Reilly is managing director for Annenberg Innovation Lab at the University of Southern California's Annenberg School for Communications & Journalism. Her research focus is children, youth, and media, and the interdisciplinary, creative learning experiences that occur through social and cultural participation with emergent technologies.

Jessica Roberts is a PhD candidate in Journalism Studies at the University of Maryland, interested in the ways in which new media are allowing for new kinds of participation and raising new ethics questions in journalism.

Miguel Sicart is Associate Professor at the IT University of Copenhagen, where he teaches game design. He received his PhD in game studies in December 2006. His research crystalized into the book, *The Ethics of Computer Games* (MIT Press, 2009). His current research focuses on developing a design framework for implementing ethical gameplay in digital games.

Linda Steiner is a professor at the University of Maryland, as well as president of the Association for Education in Journalism and Mass Communication and of the Council of Communication Associations. Since earning her PhD at the University of Illinois, she has coauthored or coedited three books and 60 journal articles and book chapters.

Bastiaan Vanacker is an assistant professor at Loyola University Chicago where he teaches and researches media ethics and law. In 2009 he published his book, *Global Medium, Local Laws: Regulating Cross-Border Cyberhate* (LFB Scholarly Publishing).

Roland Wojak just completed a MA in Philosophy from Colorado State University and also holds a BA in English and Philosophy from California State University, Long Beach. He is interested in the ethical questions that emerge from interactions between society and technology.

Sally Wyatt is programme leader of the e-Humanities Group of the Royal Netherlands Academy for Arts and Sciences and Professor of Digital Cultures in Development at Maastricht University. Her research focuses on digital inequalities and on the everyday uses of web-based technologies by people looking for health information and by scholars engaged in research.

Sokthan Yeng is an assistant professor of Philosophy at Adelphi University. Her research interests include contemporary French philosophy, feminist philosophy, and critical race theory.

General Editor: **Steve Jones**

Digital Formations is the best source for critical, well-written books about digital technologies and modern life. Books in the series break new ground by emphasizing multiple methodological and theoretical approaches to deeply probe the formation and reformation of lived experience as it is refracted through digital interaction. Each volume in **Digital Formations** pushes forward our understanding of the intersections, and corresponding implications, between digital technologies and everyday life. The series examines broad issues in realms such as digital culture, electronic commerce, law, politics and governance, gender, the Internet, race, art, health and medicine, and education. The series emphasizes critical studies in the context of emergent and existing digital technologies.

Other recent titles include:

Felicia Wu Song
Virtual Communities: Bowling Alone, Online Together

Edited by Sharon Kleinman
The Culture of Efficiency: Technology in Everyday Life

Edward Lee Lamoureux, Steven L. Baron, & Claire Stewart
Intellectual Property Law and Interactive Media: Free for a Fee

Edited by Adrienne Russell & Nabil Echchaibi
International Blogging: Identity, Politics and Networked Publics

Edited by Don Heider
Living Virtually: Researching New Worlds

Edited by Judith Burnett, Peter Senker & Kathy Walker
The Myths of Technology: Innovation and Inequality

Edited by Knut Lundby
Digital Storytelling, Mediatized Stories: Self-representations in New Media

Theresa M. Senft
Camgirls: Celebrity and Community in the Age of Social Networks

Edited by Chris Paterson & David Domingo
Making Online News: The Ethnography of New Media Production

To order other books in this series please contact our Customer Service Department:

(800) 770-LANG (within the US)
(212) 647-7706 (outside the US)
(212) 647-7707 FAX

To find out more about the series or browse a full list of titles, please visit our website:

WWW.PETERLANG.COM